'Femina Sapiens'

Patriarchy, Tyranny,
and the Origins of Sexual Psyche

Christopher Malden.

CONTENTS

PART 1   The Past

CHAPTER :-

PART II   The Present

3

PART III   The Future

INTRODUCTION

"She was soft power to his hard power; she excelled at all the roles of patronage, diplomacy and etiquette that he could not... and her gentle mien made people forget his brutality and rudeness".

'Josephine' by Kate Williams

It is not just in the present era that men and women, husbands and wives, have squabbled – sometimes raved, screamed and shouted. They have frequently resorted to physical violence. They routinely use emotional blackmail, threats, insults, cunning and duplicity.

The same couple, sniping at each other as they come under stress are, in more tranquil times, affectionate, considerate, loving, and co-operative. We live, at the best, by 'division of labour'. This term is of recent origin, from the mechanics and politics of an industrial age, yet for millennia humans have resorted to the kind of trade-offs that enable co-operative behaviour to produce outcomes of mutual benefit.

5

Mutual benefit, co-operation, trade-offs – all these terms point to the different attitudes between men and women. The differences are so profound that the recognisable face of marriage is one of compromise; living together, raising a family, juggling careers. 'Division of emotion' might join 'division of labour' as a phrase of equal significance. Modern life requires juggling emotions. Today families live on a knife-edge in what outwardly is a daily routine; what to say, when to say it, how to say it. Father's have now to be caring, kind, sympathetic, intuitive even. Mothers have to be firm, exude disciplinary power, be assertive, robust, firm, unyielding.

The following chapters attempt to unravel how 'modern life' came to be so complicated, stressful and risky. How and why it is sometimes rewarding, happy and successful. How in many instances the modern family is overwhelmed by discord, contentiousness, acrimony, resentment, bewilderment and confusion. Men feel angry and guilty. Women feel betrayed, that somehow failure of a relationship or family breakdown is inevitably her fault. 'Role reversal' conflicts with instinct.

Relationships were not always thus. The 'couple' appears to be a very recent development in human affairs, supplanting 'family' as the cornerstone of society.

Only fifty or sixty years separate the modern epoch from one in which men were required (and required themselves) to display the stern, sensible and dependable characteristics of family judge and jury; mothers ever-present, caring, sensitive, understanding, and sympathetic, comforting, and kind. Here we go back much further, to the time of the 'tribe' and before. In doing so we might discover the origins of male and female behaviour. We can perhaps discover why such deep divisions exist between the sexes; why, today, females are adopting 'male' attitudes while males are more feminine. Also, and most importantly, we discover why male and female attitudes may apparently be irreconcilable. Did males and females have distinctly different evolutionary paths, not just distinct sexual attributes? Is an attempt to coerce the sexes to combine, to become a single interchangeable unit, an outcome that benefits a 'pseudo species' but not 'us' as individuals?

These are profound questions but they can be made intelligible. Let's start with sex.

Sex in other species is a question of reproduction. In humans, the nuances of behaviour in the sexes are far more complex. For humans it isn't just sex - sex is simple only for 'animals'. There are factors in the human species that are so profound that any comparison with animals is meaningless, as we shall see. There are emotional currents and undertows, mental faculties and drives, profiles of personality and responses that define the sexes in a manner not relying on mere reproduction (which is surprisingly easy) but of social divisions and attitudes, structures that are subtle, complex, mysterious, extraordinary and, frequently, inexplicable. Then there is the complexity of learned attitudes, changing with time, different in different cultures; how do they form, and from what strange source?

To examine these entangled themes we have to confront the need for a radical re-think of the human at large. Sharing this dissection of the human condition the reader is invited to discard – sometimes wholesale – the long-accepted pathways of human development; of myth and legend, religion, much of science and other entrenched explanations of 'how we came to be'. After all, we've done it before; not long ago serious illness was caused by foul air – the best brains of the time (1800) pronounced these rigidly held beliefs even as their patients lay dying from what we know now to be illness and disease of bacterial or viral origin.

So let's take that necessary leap ahead; the reader can assess the merits of the ideas introduced here for the first time. They are simple, straightforward and easily assimilated by anyone with a lively interest in, well, life.

It is strange, then, that the crucial role of the female in 'life' is so curiously constrained and has been, apparently with few exceptions, throughout history. Females have little 'sense of self' – sometimes none at all. A friend of mine said dejectedly to my wife "I'm just somebody's mother". Sense of self requires the female to retreat into, or revert to, a confined, feminine world; having sexual appeal to the male, of softness, compliance, sympathy, insight, understanding, affection, sensuality, being maternal, endless forms that describe 'her' not 'him'.

Compare these with doctor, lawyer, builder, driver, writer, artist, scientist, butcher, minister, reporter; endless forms reflecting human roles throughout history that are essentially, inevitably seen as masculine.

At a party a short while ago I suggested to three women that despite the modern preoccupation with 'equality', women in fact had little or no basic identity.

Horror, shock, disappointment, confusion; three women of different age groups prickled. My immediate defence tactic – knowing the sensitivity and modern preoccupation with 'equality'- was to ask each their name; in full, not just their forenames, which I already knew. Their surnames were either (I pointed out) their husband's or their father's names – male names. There are no female family names.

Even going a generation back to their mother's name, even to their grandmother's maiden names, the result was the same; there are no female family names.

Unlike the male, a mysterious anonymity pervades women's lives, tacitly accepted , unnoticed, unconscious. The occasional attempt to add a hyphenated name (which may sound posh) turns into another glaring reaffirmation of the male monopoly; the name they add comes again from a masculine source.

The female, in the legal sense hardly exists, neither in the sense of a named family identity, nor as someone with their own history. There are no exceptions: even famous women, or wealthy women (bringing wealth to an impoverished but illustrious family), become anonymous within a generation or two. Only if it was to the husband's advantage on their marriage to adopt or incorporate her surname might the name survive in this new family. Sadly, though, this is just the surname of another male, the father of the bride, who gave his name (and possibly fortune) to an adored or illustrious daughter, perhaps to secure a 'good match'.

This book is not just about the many anomalies that exist in male-female relationships, but about character, personality, evolution and biology as well, as it relates to the sexes. Very hard to accomplish in this task is a neutral 'voice'; women will protest, expostulate and fume. So will men, but at different points. Because the human group – our species - consists of two very different animals.

We are not just different now, today. We have been different since pre-history. So bear with me while I 'tack', tactfully or tactlessly, along a course divergent from the turbulent wake of science and religion, of sociology and psychology. The reader will encounter new concepts that may seem at first hard to reconcile with accepted opinion, even 'proven' fact.

It turns out that widespread opinion is mostly wrong; the data science employs is often deeply flawed, as I shall show. Facts that have long been 'taken as read' are, instead, far from the truth. That they fail to 'add up' I will demonstrate. The patient reader will be content, I hope, to act like a member of a jury; consider the evidence, weigh fact against fact, before coming to a conclusion.

In doing so, it should be born in mind that mere long-acceptance of an idea or principle does nothing to confirm its truth; long-accepted ideas have their own forceful momentum, dissuading anyone to look at them afresh. 'Evolution', to use the modern shorthand, is one such idea; so momentous at its inception, so painful and contentious, so triumphant in its relevance; so coherent, so elegantly simple yet so robust. Inevitably, 'evolution' was destined to a long and successful 'run'.

But, Darwin lacked the insights scientists have today. He knew little about radiation, nothing about DNA, nothing about viruses, little about bacteria. These last, tiny microscopic, entities, are the longest-lived of any organism, a species that should more properly occupy the place 'at the top of the evolutionary tree' than we 'humans', the five or ten million-year-old 'arriviste'.

Bacteria were here on Earth perhaps four *billion* years before us. They occupy (about a couple of kilos of them) a special place in our stomachs. Without them, we would be unable to digest food properly. To suggest a really contentious notion straight away, might we not see humans as a transport system evolved to make sure bacteria access a more extensive niche?

I'm not seriously making this proposal; simply demonstrating the kind of concepts the jury might have to get their heads around in evaluating the arguments that follow, concerning what turn out to be two very different kinds of creatures; men and women.

In doing so, my intention is to simply open a door, one which before showed only chinks of light around the edges and through the keyhole. The door opens, light dawns, things fall into place.

And, yes, there is a way for women to claim their real identity (to be discussed) in a manner that distinguishes them from male identity (ditto). There is a way women could begin: select a female name from the myriad of female forenames available, one that can't possibly be adopted by a male. Here are a few examples.

Claudia, Grace, Antonia, Marianne, Ruth, Claire, Beatrice, Georgia,

Add to it a forename to give:

Jane Claudia, Elizabeth Grace, Anne Beatrice, Angela Ruth,

If you have a daughter, give her a name like this. Bestow this name now, if you have a new daughter, and from now on, in perpetuity, all her daughters will carry this as their surname; indefatigably female, feminine, womanly.

Any son you have, give the male father's name. The male line continues as it always has done except it won't be able to bestow a male name to a female. In this way, some of the tyranny of the male will end. Men have nothing to lose. Women have everything to gain - will begin once more to recapture their identity and true nature, shed their surrogacy, slough off dependency and their historical subjection. These are no longer needed, are outdated, obsolete, should not form part of a world fit for tomorrow's humans, and, as we shall see, contradict the core facts of human evolution.

In the following pages you'll find many ideas turned on their heads; how even widely accepted 'principles' are of doubtful origin. How, for example, buried deep within the pages of an obscure document in the Cambridge archives, is Darwin's Notebook 'M'. One casual entry undermines not merely evolution, his life's work, but also, unwittingly, the very conceptual base of all other sciences as well.

These are no idle claims – all will be revealed.

PART 1

The Past

CHAPTER ONE

Back to the Beginning

Two kinds of humans – male and female – have evolved in distinctive ways, with personality and psychology that seem at odds with each other. In order to unravel why we have to look closely at how humans as a species evolved. Or rather - as I have come to believe - they have evolved as a unique species along pathways quite distinct from the rest of Earth's inhabitants.

Most scientists and academics will reject this view out of hand, but I make no claim about a 'creative deity' as a reason for the 'special case' of humans. To me, the idea of a 'creator' relies simply upon a sum of qualities and uncertainties that stem from an internal mirror-image of a human-like personality. That personality is invariably male. We can lay this conundrum aside for a while. First, we have to assess this new thesis which, I hope, demonstrates conclusively that, by virtue of a weird and unique mental quirk, we are nothing like any other species.

Of course, it's patently obvious to ordinary folk that humans are different from animals; unlike us, they don't fly planes or write concertos, use mobile phones or argue about putting out the rubbish. The standard view is that humans evolved just like animals. The 'recall paradigm', central to the new hypothesis, reveals why this is just not the case.

The history of human evolution is lengthy, complex and based on finds scattered across ancient landscapes, later revealed by erosion, or from finds in caves. We have to start with the forerunners of the human family. Yet the study of human evolution is conflated with the work Charles Darwin, and his evidence for the great seminal work 'The Descent of Man' largely relied on study of the plant and animal kingdoms. My thesis is that humans departed from the evolutionary pathways he describes virtue of an obscure mutation.

This enabled humans to think their way out of the slow process of evolution as it applied to all other species. We are unique, but not through the gift of any God, rather the reverse.

Surprisingly scant evidence exists from the origins of the creature that now dominates the Earth; a few fragments of bone from sites in central Africa are the (very small, very shaky) building blocks of current evolutionary science as it applies to humans. More recent finds in deep caves in South Africa are compelling; so small are the tunnels that only women archaeologists could wriggle in and, more importantly, out. Men were smaller then, but perhaps women played a significant role in cults concerning mother Earth.

More about this fact later, and how it relates to this new theory of evolution (as it is turning out to be).

There are female bone fragments – from an early lady known affectionately as Lucy. Most fragments are indecipherable to anyone except a handful of specialists in the world, though these experts vastly outnumber the incomplete remains of the individuals representing the early forms of humankind.

So, much of the evidence for early humans – collected by the palaeoanthropologists who paradoxically out-number them – is arbitrary, debatable, contentious, scant. The evidence in question provides only tantalising clues about the developing human; about our offshoots, close cousins, poor relations, kin and forebears among a wider cast of characters; the other primates. All of the evidence of early humans is physical in nature, in the form of bones and early stone implements. Art had yet to arrive, let alone writing; another two million years had to pass by. Putting it generously, before 100,000 years – a mere eye-blink of evolutionary time – there is compete silence.

The broad picture: we – early man and woman – descended from primates, and physically were almost exactly like the chimpanzee *(pan troglodytes)*, with whom we share 98% (approximately) of DNA, the 'blue-print' that determines the characteristics of a life-form.

Darwin and his generation were unaware of this; unaware both of this precise figure, or the existence of DNA. Furthermore, the mere idea of such an association, however slight, caused outrage. For some humans, it still does.

Other proto-humans, quite a few of them, became extinct. We know just a little About them; bone fragments, jaws, teeth, portions of skull, represent (going backwards through time) *Homo neanderthalenthis, H.heildelbergensis, H.Antecessor H.erectus, H.Ergaster, H.habilis, H.rudolfensis.* Each name is a bit of a mouthful, again greater even than the actual number of teeth that constitute the evidence.

For *H.antecessor* and *H.rudolfensis,* in particular, there is insufficient evidence to indicate size of the brain cavity, etc., merely enough for them to be placed along the human evolutionary trail stretching out 2 – 3 million years that eventually led to us.

All of them, save perhaps *H.neanderthalensis and H.floreseniensis,* inhabited earth for a far longer period than present-day humans. It's a sobering thought that a 'primitive hominid', in the shape of *H.habilis* survived very effectively over a period between 3 million and 2 million years ago. A million years of history, or more. Modern humans account for just a thin smudge of time; about160 thousand years. It's worth bearing in mind the names of these ancient creatures are invented by modern humans; there was no one around to consciously calculate, observe or name them.

The shark meanwhile, and it's a long, long while, has lived in the oceans for around 350,000,000 (yes, that's millions) years, also an ancient ancestor of man.

Pre-dating the shark family, another ancestor – an extinct fish called the placoderm – was the first to enjoy that feisty jousting ground so beloved of today's humans – sexual intercourse [1]. So that's been going on – a juicy snippet for any social gathering – for 375,000,000 years, pre-dating sharks. By now no one should have any excuse but to be an absolute expert.

Imprecise time-scales bedevil our attempts to construct accurate human history. Spanning a period lasting around 6.5 million years, an array of similar candidates for our primate ancestors present themselves in the form of tiny scraps of bone. In other words, a lot of forensic guess-work goes on.

Paranthropus bosei for example, lived through a period of over a million years. This long-dead proto-human, dying out around 1.2 million years ago, had functioning bipedal gait, a massive set of teeth but a very small brain. Not someone you would want to meet on a dark, rainy night.

'Niche' is an important word that will recur with some frequency. The concept of the niche is vital to the story of the earth and its inhabitants. It is important to state also, early on in this peculiar tale, that there are perhaps 8 – 10 million other species on earth, apart from present-day humans. Each species consists of millions of millions or billions of billions of individuals. This is the true population of the Earth. So evolution isn't – and never was- just a question of human evolution. See: James A. Long, Scientific American, Jan 2011.

We do know, from the evidence, that *P. Boisie. H.rudolfensis, H.habilis* and *H. Ergaster* all co-habited within a relatively small region in northern Kenya around Lake Turkana. How they interacted is unknown, though we can be almost sure they knew of each other as competitors for the same kind of food; they shared the same 'niche'. Perhaps the males raped females of a similar type –but this is a subject unspoken of (as far as I know) within the scientific community – a sensitive bunch, also inhabiting a rather specialised 'niche'.

Of all the species that *ever* lived on Earth 99% of them are extinct. These might total (the figures are tentative) to five billion species that we'll never see again. And of the current total of 8-10 million *different* species, perhaps 86% are, as yet, not even fully described (someone you know, your nearest and dearest, or a son or daughter, you might feel also may belong to this group).

We are one species among a vast multitude. We are outnumbered billions to one. No one has counted; each species occupies its own niche, each niche in turn interdependent upon all others, many overlapping.

So, when you next turn on the TV to another science programme, bear in mind – as the hushed, reverential tones of the voice-over extols the extraordinary power and insight of the scientist-cum-high priest of hallowed knowledge – that in fact we know very little. We know almost nothing about nature or its past, about humans and our forebears. Most of what we know is highly speculative and quite likely to be overturned any day now. The time scales involved make our supposed understanding look wildly optimistic, not to say presumptuous.

Of our two (possibly) closest ancestors *H.neanderthalensis* and *H.floresiensis,* an important key to comparisons in terms of human development – before we even get to males and females – is to remember that these, our most recent living ancestors, were very remote from one another . Remains have been unearthed on the peninsular that is now Gibraltar and in Indonesia. These niches were both distant and different, thus diminishing competition and conflict.

H.floresiensis became extinct 'only' around 18,000 years ago. Surprising small, and despite a childlike body and skull size, *floresiensis* perhaps had a lively and vivacious brain – we don't know.

One vital piece of the jigsaw that may be crucial in the attempt to delineate male and female personalities in the modern human is to know how far pre-modern humans may have 'interbred' (to put it delicately) with each other.

We (or rather, scientists) are pretty certain that *H.Sapiens* (us) and *H.Neanderthalensis* did interbreed and, in doing so, of course, passed on characteristics – mixed their DNA. Being remote at Flores in Indonesia, *H.floresiensis* perhaps did not, even though they are also 'contemporaries'. Perhaps it's a pity we didn't 'know' them as well.

If we could successfully interbreed with close hominid forebears, we may find evidence of shared characteristics between *H.sapiens* and *H. Erectus*. This latter was a highly successful group, spanning a period longer than any other, from 2 million years ago (or thereabouts) to the very recent past, 50,000 years ago

At the beginning of this 'reign' of dominant hominids, the human niche was shared by four others, among them *H.habilis*, the first hominid associated with tool use and so is a prime candidate for having 'passed on' this essential human characteristic.

So we arrive at a crucial point; this is the sole, seminal, piece of evidence that points to a unique event in evolution, at least as far as humans are concerned. For evidence of tool-use suggests something beyond what even Darwin (together with most modern scientists) comprehend as 'evolution'.

For with the appearance of tools, evolution itself evolved. Many a palaeoanthropologist will baulk at this statement. But, with the appearance of tools we left all other animals behind. This was the point at which humans stepped outside the evolutionary pathways whereby other species develop – the evolutionary pathway that Darwin so cogently described.

But while all other species dutifully continue along Darwin's evolutionary pathway, humans did not. Times change; not only do species evolve, evolution has evolved too.

By enabling us – I can call us 'us' now – to modify our niche, even these primitive tools demonstrate how we were no longer reliant upon the slow passage of time (cornerstone of Darwin's theory), to adapt to the environment. Instead, we adapted the environment to *our* needs.

Evolutionary science has missed this crucial step and the theory of evolution should really now be re-written. As you read this it might be happening. To date, evolutionary scientists have accepted that the term 'tools' or 'tool use' means either bits of stone or wood employed in feeding or for use in showing aggression or in play. Thus whole tribes of animals, birds, bugs and primates are defined as 'tool users' thereby conflating human and animal 'tool use'.

Here we make a distinction for the first time between different types of tools. One type we can call 'extemporary' tools; a naturally occurring 'piece of the environment' that serves as temporary expedient (using a twig or thorn to dig an insect out of a crack in a branch) and then discarded. Humans employ what we might call 'manifest' tools; they are envisaged for a purpose not yet present.

This crucial distinction dispenses finally with a useless need to confuse animals with humans. A gradient of capability or intelligence that includes all animal and plants does not exist. The imaginary graphic representation starting at zero and escalating – as Darwin believed – to the dizzy heights of that ultimate expression of 'life', the white, human male, is entirely false. The most successful life form ever, 4 billion years ago as well as today, is bacteria.

Yes, there is intelligence in apes, just as there is intelligence in spiders. It is ape-like intelligence and spider-like intelligence, neither of which we can measure adequately, emulate or understand unless somehow we can inhabit their heads, bodies and their niche.

Thus, at some time in the dim and distant past (a cliché, but startling in its accuracy) humans 'invented' manifest tools and in doing so diverged from the evolutionary pathways of all other species. Other species did not invent tools to so modify their environment that the niche itself began to evolve.

So far, these concepts appear to have eluded evolutionary scientists but the stark fact is that humans now inhabit a mental niche that expands exponentially under the influence of reciprocal input from both niche and its inhabitants; we read, and are prompted to write.

We find materials and change them, adding them to a stock of implements and resources not until then present in nature. This is not the way any other animal behaves. Behaviourists will excitedly point to 'anvils' (or to the 'favourite stones' of sea otters) used habitually to break open shells. These are not 'manifest'; they are not 'made' in the sense of preconceived, imagined or fashioned. These are concepts that apply uniquely to humans.

Evolutionary science has overlooked, I believe, these important distinctions and the crucial evolutionary step that pre-dated their emergence. In the next chapter we'll see why some crucial re-writing of natural history will need to be done to exclude humans from the accepted notion of a convenient gradient that compares 'primitive' with 'advanced' or 'instinctive' with 'intelligent'. Humans, it seems, no longer belong on the same trajectory, to the same list of species or even to the same evolutionary 'page'.

And so we come to another milestone that collides with the wheel of the apple-cart and so overturns it: we have to ask another crucial question; 'Did women invent the first tools?'

This 'cultural shock' of a question is similar to that which we experience when it is pointed out that women don't even possess their own surname: Or when we question the tired, science-fantasy of hairy blokes in ripped animal skins charging around after mammoths.

These interpretations are no longer representative of the kind of searching-after-truth that pays due respect to today's men and women. Especially women, and in this sense evolutionary science fails. We are intelligent creatures. We understand common sense ideas. So we ought to employ them when it comes to the question of our own evolution.

Stone artefacts last millennia. Large stones were, of course, very useful; apes and monkeys use them as 'hammers' and 'anvils'. These are the durable evidence from early human-like activity. Scientists, in their wisdom, speak of 'early man' and make assumptions accordingly. Females are elided from history because no one has given them a thought, let alone crediting them with the invention and first use of tools, or allowing them a name that can be passed on to female offspring.

There is no direct evidence that males invented the 'first' tools. There is only supposition. But we can State with absolute certainty that *of course* women of early epochs played a significant role. Women are adept, insightful, and conscientious. Skill does not require muscle power. Any mental advance would inevitably involve all those who shared in its advantages and benefits, women foremost among them.

Instead of stone weapons, let's look at the possibility (using common sense, the kind we've inherited from ancient ancestors) that the first tools were not stone, but wood, horn and bone. This is entirely in accord with observed use of 'ex-temporary' tools as used by animals; twigs, pointed sticks, sharp thorns, branches.

It's no great leap – indeed it's a very tiny step – to imagine a woman using a thin, pointed stick to tease out an edible grub or impale a small bird, mouse or fish for the purpose of roasting it over the embers of a fire. Yes, the barbecue has a long and illustrious history.

When archaeologists speak of 'the first tools' what they actually mean is 'the first identifiable tools (stone or flint, by their shape or by their cut marks on recovered bone) that have survived over a period of perhaps 3.5 million years'.

Glaring omissions from the assembled evidence is the unspoken fact of a) women b) barbecues c) sharp, thin wooden sticks. This analysis may also include: piercing leather, threading thongs by means of hooked ends, longer spear-like fishing implements, basic cooking utensils, grubbing tools for roots, plants or insects, pulling down fruit or nests by means of straight or forked branches. The possibilities are endless, but the implements don't survive.

Because they have rotted away (understandable after 3 or 4 million years) this does not mean – through lack of evidence *in situ* – that these tools, the processes and the women who may have invented and used them, did not exist. The evidence is, to the contrary, *a priori,* precise and compelling: women imparted DNA right down to modern humans. Women today use exactly these implements and tools; a nice brochette of grilled *coquille St.Jacques* flavoured with peppers and onions, for example.

Thus, tools fashioned from stone were far more likely to have emerged much later. As the technology migrated to enhance human effectiveness in more elaborate forms of food gathering, like the hunting party, it is highly probable that stone tools were derived from their wooden precursors, that women were not only familiar with their originals but had a hand in their manufacture and use. The implication that males invented the first tools is lacking in evidence and may well be false. The falsehood may not have been stated, but is always implied.

We are speaking of a time when human life was very like primate life. Were females dominant then, heading up a hierarchy of well-bred *H.habilis* while the males competed for the attention of the females? Did the males strut around, demonstrating their finer points of strength, size and agility to gain perhaps the privilege of mating with the prime example of her sex?

To gain insight into these questions is to abandon many assumptions about how one single emergent species (erroneously called 'man') came to dominate the planet. We have to abandon the premise of males leading the way. This idea came later with the dominance of patriarchal social systems that relied upon Biblical and other ancient texts, thereby giving rise to the myth of a male God, a male Christ, Buddha, or Mahomet, among others. Could it be that this later prejudice may have unconsciously coloured the view and distorted the judgement of those whose specialist subject is pre-history. Does it still?

These later patriarchal epochs thrived via complex social organisation and a convenient conspiracy – convenient to both sexes – in which males competed in warfare and slaughtered others in the competition for already scarce resources as they became ever scarcer. In return their women and children, though shielded from the brutal realities often instigated by men (not necessarily nature) were a mere supporting cast.

Journeying even further back we can well imagine (though there is scant evidence) a time closer to the life of competing creatures less vicious but just as hungry and determined. Weapons, quite literally, gave us an edge. But the first tool may just as well have sprung from a female; to suppose otherwise constitutes gross intellectual misconduct.

Mutation, we can perhaps at least agree, is the probable cause of our surprisingly sudden emergence as *the* dominant species. Evolution is the stately progression towards stasis, the process of the Darwinian imagination. Mutation puzzled Darwin; he was familiar with the work of pigeon fanciers (he was one himself) but never successfully explained the rapid changes brought about by selection that was anything but 'natural'. Human breeders were choosing the traits they valued. In stock rearing this practice was vital to the success of English agriculture, and still is today.

By confining his arguments and observations to species which conformed to its precepts – a process still adhered to by evolutionary scientists – Darwin avoided a curious paradox; humans were not just observers and descriptors of evolution, they were among its prime movers and agents.

More especially, his principal theory is confined to the way it affects *other* species, not humans and their mad, headlong rush for progress, expansion and new ways to shape nature itself.

So humans could not be an outcome of evolution in the classical, Darwinian sense. Instead, the emergence of a 'pseudo species' defies scientific description: it refuses to obey the rules, creates its own niche, exists outside the natural boundaries and the confines of the Darwinian view – is inexplicable unless we accept a new paradigm.

Quite contrary to the laws of natural selection, humans continue, at an exponential rate, the construction of an ever-more complex artificial niche – a mental niche – that has no relation to anything that Darwin (or anyone else) could classify as 'natural'. And Science is mute.

Thus, the persistent chauvinism of the modern scientific community has its roots in a flawed theory, apposite for describing species in their natural element but lacking any parameters that adequately encompass 'man'.

Let's ignore or by-pass the chauvinism and say a woman, perhaps early in her life, was the site of a mutation.

Good candidates for these mutation events are sudden excess of radiation, the release of toxic chemicals, perhaps the aftermath of volcanism, excessive stress producing high concentrations of cortisol within the placenta. It's unlikely we'll ever know. But to realise what was a stake, let's take a look at the female genome.

Hardly different from the male, though equally likely as a candidate to account for the sudden emergence of a pseudo species, it had to undergo subtle but cataclysmic genetic change. Like the switching of the points on a railway line, the whole genome was quite suddenly , in a very un-Darwinian manner, diverted towards an unforeseen, unplanned, un-knowable destination.

Small changes at critical junctures have disproportionately large effects on future development. In the case of humans, a mutation affecting mental architecture would – actually did – produce such a disproportionate effect, the one Darwin overlooked. [3]

Paradoxically the future of all species on earth, was written into the human consciousness.

[3] Well, not quite; subsequent to the publication of a previous work ' Dangerous Mind – On the Origin of Pseudo Species'(2006) which first sets out the special role of recall in human development, I discovered, that Darwin makes a passing reference to its crucial role. This will be discussed in a later chapter.

CHAPTER 2

Memory and Recall

A discrete mutation might have happened to a female hominid, *must* have happened to a human ancestor (possibly to a female), prior to 3.4 million years ago.

These dates are precise; the Earth's magnetic field reverses from time to time. This magnetic signature, imparted to rocks, sediments and clays, provides accurate dating evidence for finds embedded within them.

Lucy, the lady we met before, was an *australopithecine,* an antecedent of *H.habilis* (originating around 2.8 million years before the present). Lucy existed 3.3 million years ago. Cut marks on animal bones found in deposits 3.4 million years old were made by tools manipulated by a distant ancestor. These findings were the subject of dispute (perhaps movement of stones in a torrent caused damage?).

Today we have better evidence of the ancient nature of human activity, consciousness and thought – for this is what tools tell us.

Sonia Harmand of Stony Brooks University, New York, got lost in Africa. She was looking for a site near Lake Turkana and took a wrong turn. She and her co-workers stumbled on a find that comes close to confirming this earliest use of tools.

Lake Turkana of course was the site where the remains of our ancient friend Lucy first came to light. Close by, in terms of distances in Africa, Sonia Harmand and her team found knapped flint tools and flakes from their manufacture firmly embedded in deposits securely dated to 3.3 million years ago. Even the flints carried the palaeomagnetic time signature.

It's a stretch of the romantic imagination to suggest Lucy made tools like these. Yet the tools and Lucy (her remains) coincide archaeologically with uncanny and unaccustomed precision.

Is it such a stretch of the imagination? Well, let's try to stretch our imagination It can do no harm. Women are dextrous, methodical, patient, precise, persistent, consistent and as bright as men. Surely no one could disagree?

These are the reasons why they were employed in later industrial contexts – sought after by employers – in which repetition, dexterity and precision were valued qualities in a skilled worker.

We have to also examine the tools themselves and see what they're like. The first quality they possess is that they embody the nature of memory. This is why I emphasize what *must* have happened; memory is essential for tool manufacture. The prior condition necessary is only that the mind grasps the flow of events, their causal relation. Recall prompts the realisation of what is required for a more beneficial outcome: this new mental 'app' augments the present with a picture of the future. This powerful tool re-shaped evolution; a mutation that changed the course of the history of planet Earth.

With recall humans 'leapfrogged' the evolutionary process, sidestepping the unbelievably long passages of time required for species to change .

The ability to recall events 'at will' confers an unquestionable advantage – is the quintessential precursor for tool manufacture and use. This facility is unique to humans and unique in nature.

Scientists will again bristle and splutter into their cocoa. However, I await with interest any possible alternative explanation for the dominant position of humans in the earth's biosphere; we are puny, physical underdeveloped, not very robust and lacking stamina. We have relatively poor eyesight, no real sense of smell, feeble teeth, no claws, thick skin or warming fur.

Yet all animals go in fear of this strange, unique, upright (bipedal), skinny, puny, hugely ubiquitous, fearless newcomer armed to the teeth with an alarming array of inert but inexhaustibly powerful weapons.

One crucial mutation caused Evolution, that stately natural process capable so elegantly of delicately shaping such an infinite variety of myriad species on Earth, to swerve sharply off course and produce, in evolutionary terms, a monster.

Recall sets us apart now, just as it did in Lucy's time. A sharpened flint did not change evolution into revolution. Recall sets in chain modes of behaviour that defeat all other contestants in the competition for a suitable niche. Suddenly – 3.8 million years ago – life became a casino with the odds stacked against long-established species. In the game of life, the tables were rigged in favour of humans.

The hallowed evolutionary process once imagined to be sacrosanct or immutable is, it turns out (if this new theory is correct), a process carrying flaws of its own. Evolution *itself* evolves, perhaps. Worse, it may be subject to its own inherent mutations and flaws.

On this gloomy note we return to the present; nothing really of note has happened, after all, since Lucy and the advent of recall. This latter describes adequately all subsequent developments of hominines and their all too obvious impact on the world's other species; these latter have declined in number variety and significance, humans have come to dominate all other species.

And in the present or, more accurately, the recent past, we encounter the strange emergence of two types of human; males and females, which we must explain.

Time to return to our original concept: that women are, or might feel themselves, captives when they ally themselves to a man. Historically, women supposedly gained the better half of a bargain; a hard-working man would provide for her and her children. In return, a type of voluntary slavery, rather than the dreaded possibility of bleak sense of failure, emerged.

Were men any more fortunate? Historically their role as provider and of fathering children enshrined male status. The relationship might provide even more – perhaps greater freedom, once their (brief – sometimes all too brief) sexual partnership was established. The woman, now mother, would look after 'his' offspring. He could concentrate on the vital business of work or career, building the foundations for a better future. Yet this would include long-term responsibility and a need to apportion more of his time and effort to include a family growing in need – would he become a hostage?

Two scenarios point to different perspectives; are they old-fashioned, outmoded? Have humans progressed beyond this 'division of labour' the term adopted from popular economic jargon?

The personalities of male and female are so deeply imbedded that the answer is probably 'no'. The male has a toolbox – even if it's under the stairs, the kitchen sink, in the boot of the (or is it 'his'?) car, in the garage or garden shed. No matter what the material circumstances of the family, the male has a deep affinity with tools.

Women (of the more traditional aspect) might have a box, too. But it's called a 'work box', is smaller, filled with precision instruments; pins, needles, cotton threads, scissors. Another for pens, envelopes, writing paper, stapler, paperclips. We haven't visited the kitchen yet.

The kitchen is a treasure trove of wood, metal, glass, steel, ceramics, electric devices and an array of impressive machinery: fridges, vacuum cleaners, cookers, grills, microwaves, ovens, mixers, blenders, shredders, sieves, can-openers, corkscrews. Still we haven't visited the bathroom, briefcase or the handbag. We don't need to, in the bedroom, the drawers of the female's tool bench (otherwise known quaintly as a vanity unit) are curlers, tongs, bending tools for eyelashes – this author doesn't know what they're called – rollers, grips/clips, mirrors, buffers, clippers, nail files, heel smoothers, strips of wax, tweezers. We haven't mentioned products, like lipstick, eye make-up, powder and the implements needed to apply them

Neither is this cursory glance at the 'kit', which even the most 'emancipated' woman of today regards as essential, in any way denigrating; it forms the basis for an argument that again attempts to place women at the centre of an emerging species called (perhaps mistakenly) 'man'.

Even Google now issues a red alert (electronically speaking) when I write 'man' in the connection of the human species; with a deprecatory cough, the word is underlined and I'm nudged into accepting the more neutral 'humankind'.

But we still speak of 'Man' as opposed to 'Beast', though perhaps this distinction has been blurred for some time. So our examination of the female 'toolkit' is relevant, especially when we find that the male version, while being considerably more chunky, heavily-built and sometimes obsessively worshipped – perhaps enshrined in the garden shed impaled on a row of nails – is not nearly as specialised, as frequently put to use, more closely associated with the persona.

A man might go for weeks without getting out the toolbox, often at the very time when the dripping tap needs urgent attention. Female tools, meanwhile, are pushed to the limit in a relentless round of servicing, repair, adjustment, experimentation.

This laboured point leads us to examine further the inherently narrow view of evolutionary scientists. It reveals the strange hangover from male-orientated patriarchal lines of thought that nothing in the 19th, 20th or 21st centuries has accomplished by way of embracing the so-called 'enlightenment' which Darwin's theory of Evolution by Natural Selection supposedly ushered in.

To many at the sharp end of investigating human evolution it is unthinkable – unacceptable – to suggest that the first known tools were the work of women. It is as if one asked a Christian scholar why Mary, the wife of Christ, hadn't produced a girl child instead. These are fundamental questions, like the question as to why there are no female family names. And which is why we have to resort to shock tactics to bring the question of human evolution sharply into focus.

There's more trouble ahead.

First, the literature of evolution mentions nowhere the undisputable fact that recall must be the prime mover in the rapid progress of humans up through the ranks of other species. There are about 1.8 million other distinct species. Only humans employ tools. Many researchers argue that many animals use tools.

This is true, until we make a distinction between two types of tools; those that result from a close environmental stimulus (let's call them extemporary tools) and human tools, which are a permanent fixture of our special 'niche' (too permanent, when it comes to taps maybe) Of the special human 'niche', more about this later.

The only extemporary tool that's even close to that of the human variety is the type I've seen used by sea otters along the Pacific coast (in this case, north of San Francisco at Carmel). Here, as with sea otters elsewhere, they feed on molluscs using hard stones as anvils to crack the shell of their mollusc meal. Balanced on their chests is an anvil while they swim on their backs. The stone anvil is 'theirs' and they may favour one stone over quite a time.

Other tools of this specialised kind appear in a few other species; the orang-utan is able to fashion leaves for a particular task, depending on the difficulty of teasing out an insect from a thorny hiding place, a grassy one or a fissure in a tree. He or she can also fashion a cap or canopy to keep off the rain. But, these are discarded and must be made again. No problem, orang-utans are very clever orang-utans.

There is a distinct wish by animal behaviourists – it deserves to be a field of study itself – whereby the object of research is to prove levels of intelligence comparable to that of the human are present in animals. This is redundant; scrub jays are intelligent scrub jays, elephants are sensitive, empathetic, very intelligent elephants. They are not human.

The human use of tools, the human gamut of intelligence has nothing to do with animals. What animals have, we have lost, and cannot regain. There is a sense of loss, perhaps, a wish to redress the balance, by making animals, for which we have enormous affection, respect and regard, part of our family. We cannot and, already domesticated animals aside, it is a crime to try.

Indeed, it is human intelligence and how it translates into tools and tool use (in the broadest sense) which next we have to confront.

I have forcibly claimed that recall is the precursor of our way of manipulating the universe to suit us, a strategy so elaborate and effective that it far outstrips the efforts of all other species. When it comes to competition, we out-compete all-comers, even lethal bacteria and viruses (at least, so far).

Our ability to recall events at will is the basis for imagining that the world, nature, weather, life, slides past us; we have knowledge, through recall, of what we call the past. Our experience of how events unfold permits us to anticipate – we have a pretty good idea of what's going to happen next.

This scheme – unique in nature – whereby a virtual world is 'smeared out' into past, present and future, has no real existence in fact. It is a uniquely human construct. Which is why no gorilla, nor even a crocodile, has ever been seen with a timepiece.

Humans live an illusion – there is no past or future, only recall and anticipation. But this illusion is so powerful (and so successful –it got us where we are today) that we can hardly think these thoughts. It takes a quantum theory specialist to confirm that, when we examine the flow of events in fine detail, we can't trap 'time' or tie it into a coherent physical entity. We can't examine time past, except in history books films or photographs, which exist in the present. We can't go into the future, that doesn't exist anywhere, either.

This is the world other species live in. True, some are habituated – they get used to drinking in the same pool, roaming the same territory. They also may have great difficulty assimilating anything 'new'. They may take no notice at all, depending on the class of animals to which they belong. An ant will crawl unquestioningly up the side of an aircraft carrier.

Humans will fit any new object, event, and experience into a stored yet easily accessible hierarchy. By adulthood, the hierarchy is vast in comparison to more specialised species. We can safely say of species that many of the seeming 'advanced', 'intelligent', 'complex', 'responsive' varieties are almost inevitably also 'endangered'.

This sad fact stems from our own intelligence, the kind that can scroll back and forth in virtual time to plan future action, learn from a vast store of experience whereby we can easily recall which response was most effective 'last time' (even though there *is* no such thing as 'time').

I did mention there was trouble ahead. But, now we go back to Lucy and the possible first use of tools. How this event leads to almost separate male and female types, how the persistent action of experience becomes engraved on our consciousness, and how – more than 4 million years since its possible inception – the human mind alone begins to make its inevitable, eradicable mark upon the biosphere.

So while evolutionary scientists will doubtless reject the idea that women made the first tools, even though they have not the slightest shred of evidence to indicate males were responsible, we do have the evidence that tools are very old. Palaeomagnetic values give signatures in strata, tool core and flakes that date early flint tools to 3.3 million years ago.

This is not the date recall first presented itself in the brain of the australopithecine (perhaps a female). That came earlier. Used first in a manner that left no trace, in planning behaviour that would do most to ensure success in a hunt for game, more likely recall is most useful when searching for birds' eggs or the roots of plants known to be filling. Recognising edible fruits, distinguishing between poisonous and non-poisonous tubers, leaves and roots, knowing where food is easy to reach, carry and prepare needs precise and effective recall. It enabled early men and women to repeat previously successful strategies. Unlike animals, recall is independent from environmental stimulus. In other words, the place needn't be the same, not the time. Even without environmental markers, recalled strategy applies. Consciousness begins to replace instinctive response.

It is in this way, long before the use of tools, that recall must have shaped behaviour to allow the primitive biped to outwit competitors from less well-equipped species. Recall provides a novel ability to tackle new territory, new and different ranges of food and terrain. Recall is a repository of data that is versatile and adaptive to a larger set of encountered circumstances, a facility quite unlike instinct.

How does this relate to Lucy and her sisters? At this juncture we have to make a case, again, in part to answer the inevitable protest from established evolutionary theory.

In a sense, this must have been a golden age in prehistory; proto-humans were in no position yet to displace other species from any position of dominance. They had a slowly developing consciousness prior to the advent of tools that gave an advantage yet to fully emerge. The slow process of a new ability spreading through population would have taken hundreds of decades. But the crude caricatures of the science magazines, of brutal looking males in conflict with ravening beasts, battling the environment, fighting competitors to the death for the sake of a square meal is most likely fantasy.

Strategy far more sophisticated than in any other creature yet evolved was now present in a single species destined for domination; the human. Now humans were developing a new way of looking at the world, independent of instinct, strategy meant primarily protecting valuable individuals, keeping out of harm's way, safety first, not taking on the world. The ferocious hairy hunter image arises from a mistaken view of early man; women were not only there but clever, too. This was undoubtedly, prior to the dominance of flint, the world of *Femina sapiens*. Emphasis of patriarchy at this early date stems from a world of 'science fantasy'.

That planning 'paid off' was an early learning experience that emphasised the need to hone abstract mentality.. Recall provided foreknowledge of where and when food supplies might be plentiful. It helped in assessing new territory compared to that already known. Recall encouraged subtler mental responses; looking out for the telltale signs, anticipating danger. It was safer.

Instead of relying on chance encounters with food, water and weather, recall had other advantages – it helped save time, distance, haphazard wandering. Recall paid off for Lucy and the other women and men in her clan.

Even at that early stage, recall would prove to be a considerable advantage. Recalling routes back to water or to a safe resting place, aided by simple blazes and marks. A method still in use today, the blaze is a crucial, defining quality of recall, emphasizing its abstract character in Nature, hitherto lacking any abstract elements. No other animal can 'blaze a trail' employing the subtle, abstract markers rich with a significance unique to humans.

Thus, devices uniquely human may easily pre-date the tools we find. The simple device of the blaze or a sign, meaningless to animals, meaningless in Nature, is the defining mark of recall and the sign of the human. Try as we might, animals still fail to recognise road signs.

The brutal world of pre-history as depicted in even apparently serious journals is, I submit, comic-book material. It is unlikely to be the case that, as hominines progressed they eliminated competition by physical force. It's more probable that other species simply failed.

Nature has 'produced' large numbers of animal and plant species; test beds, as it were, of physical types, which over a certain time scale can successfully exploit a particular niche. However, a niche can change and a key advantage is adaptability. A species able to exploit many different niche environments has a great advantage. Reliance upon a specialised niche that is vulnerable to natural disruption often spells disaster, perhaps extinction. The dinosaur and the shrew demonstrate the case.

There's a script we all follow written in our genes; DNA. Many species have no choice but to fulfil their prescribed role. Our DNA, with DNA written into the script we follow, allows anticipation and reflection. That crucial piece of script makes us human, not chimp, crocodile or beetle. Chimps have their script. 98% of the letters are the same for a modern human. Very few – but crucial letters – creates a difference so significant, that whether they are in the wild or in a zoo, they depend on humans for their long–term survival.

Of course, the above is a simplification. The human we are speaking of (female or male) had a genome consisting of 3,400,000,000 components – the number of nucleotides; the bits that may be imagined as keys on a vast piano and the genome as the composition. Hidden in the manuscript are sequences of genes that form the code for proteins. There are around 25,000 of these spread out over the 23 chromosomes that our female ancestor possessed. These she passed on her to offspring, passing on similarities relevant to the human species.

Alongside the legacy that constituted her form and personality, appearance and overall character, the composition includes in the vast sequences of nucleotides: memories of her ancestors and the more primitive states of nature she once inhabited. Her version of the sequence was more similar to the chimp than it was to a fish but, curiously, they are all there. All species have descended from previous versions of the same composition, small variations here and there.

Most of the work the genome does is to preserve what we are, because, if we survive and breed successfully, give birth to offspring with the propensity also to be successful, the species survives.

That is Darwin's insight; here is no master 'plan'. Thus, over aeons of evolutionary time, we see, in species alive today, the same system, the same components, but with slightly different profiles, able to exploit the different niches available in the physical environment. The multiplicity of possible niches determines the variety as well as the number of different species

Some niches are other animals; like the bacteria in our gut, the fleas on the cat. In humans, though, the HAR1 gene and the FOXP2 are distinctive and significant. We tend to stand out from other species because of coding for big brains – though a recent ancestor, *H.floresiensis* which, in spite of tiny stature, was still apparently an equal contender for further advancement, stumbled unaccountably at one of nature's hurdles. A very recent and perhaps close human cousin disappeared, forever.

Nevertheless, we have 25,000 genes with around 1,500 nucleotides each which represents a tiny portion of the 3.4 billion nucleotides; just 1 or 2 % translates into making us. What does it all do? Perhaps nothing at all, or is it the product of natural catastrophe?

Darwin would probably say 'yes'! Our female with a new, minute, but significant change to her genome might have the propensity for great things, if the opportunity presents itself. But if disaster strikes, the genome has so many components that, if the clock suddenly begins to run backwards, components are available that might stand a chance. Life is nothing but indomitable. So is death or extinction; errors are eliminated by these simple expedients.

Modern research now concedes that the other 98% of the genome, seemingly passive before, is actually highly active, in regulatory processes not yet understood. But we can be sure there is an evolutionary state that is active, not passive, perhaps 'geeing-up' the genome continually, ready to grab the next mutation that will help the species adapt as rapidly as possible in a fast-changing environment.

CHAPTER 3

The Hunter and the Gatherer

The patient reader may now be more alive to the fact that the history of mankind (sorry, human-kind), written by inheritors of a patriarchal mindset, is awash with exaggerations, clichés and questionable science (more of this later).

One cliché is notorious; the concept of the 'hunter-gatherer'. Even in the mind of the anthropologist and evolutionary scientist, over the idea of early man hovers an image straight out of a bad illustration, an ill-informed artist's impression of an earlier world. No doubt persistent attempts to 'popularise' the subject have conjured the image of both 'man'; ragged, hairy, savage, muscled, a battling figure conquering nature and ferocious beats.

The imagery seems to have influenced even the professionals: women have merited little attention as perhaps progenitors of human behaviour and skills.

In the on-going attempt to 'popularise' subjects is science in some way perpetrating a fraud? Has science handed over to artists and film-makers the responsibility of interpreting an understanding of early human development that may easily be wrong? Probably 'Yes'.

Let's consider one such vision that might be just as believable, perhaps even closer to the truth. Hunter-gatherers, a group of picturesque primitives, women trailing after men, children trailing after women, old folk trying to keep up, a sturdy male, hand shielding his eyes as he scans the far horizon and a group of heavily-armed figures, some carrying the ubiquitous hefty stone axe head, bound by leather thongs to a hefty stick? Probably inaccurate.

Instead, there are two pictures which, taken together, make much more sense. One is of a cautious, furtive group of males. They move slowly, are quiet for long periods, they communicate by subtle gestures. They carry slender spears and spear-throwers. They are not hugely muscled but more like runners, built for stamina as well as stealth. The prey is now surrounded; the hunt enters a new, more deadly phase. They prey is not a fearsome, savage, thirteen-foot-tall monster covered in masses of hair and sporting improbably large tusks. Instead, a small fawn or a small group of quail is nervously scanning the scattered shrubs and trees, looking for a way to escape these weird predators.

In the other picture – a long way off from the hunters – are the gatherers: these are fit, sturdy women. They scan the ground, bushes and trees, with keen, accustomed eyes. They have hair that is short or tied back out of the way. It is immediately apparent they have an eye for detail, a demeanour suggesting deep concentration, experience and knowledge; they know what they are looking for.

Youngsters accompany them, watching from a respectful distance and carrying finely-woven bags of plaited fibre of reed, grass or strips of thin plaited tree-bark. They wear protective clothing; their legs and knees bound with animal skin to ward off thorny scrub, coarse grass and damage when they kneel. They carry carefully crafted digging sticks, pointed and hardened by fire, forked sticks and crooks for pulling down branches. In the basket are tubers, rodents, snakes and other small prey trapped or prised out of holes, birds eggs, fruit, berries, tubers, roots, leaves, indeed anything edible. Older women and men may tag along, offering advice, for this is a skilled and valued labour.

The main point here is that hunters and gatherers are two distinct groups, each with a separate set of skills. It is from here, from this far-from-primitive division of labour, that the crucially different personality and psychology of the male and female arose. At least, this is what I believe. It is still present in current tribal behaviour. Women do not trail behind men or adopt subservient roles.

The male, from the time of this specialised activity, prompted by the conscious fabrication of tools, and the forethought involved in planning a hunt, shaped his personality. We can see it today; linear logic determines his personality. Understand this and you understand men.

Men identify an objective, they commit themselves wholeheartedly and without reserve to the task in hand. Formerly, the risks involved were great; not that of death necessarily, through injury inflicted by some huge beast, but by failure. Failure to 'bring home the bacon' meant the perils of hunger in a harsh world would loom ever closer. There would be some grim silences around the fireside.

Linear logic becomes a successful mental approach to maximising return (food) for the effort expended. Ask any hunter (and I have) what it's really like to scale a couple of mountain ridges and come back empty handed. The reason for failure is not usually found in the linear logic but in human failure; someone went the wrong way or didn't stay close enough to his left-hand or right-hand companion. For this is the way men tackle tasks, it's the way hunters hunt. Linear logic is abstract, has a hierarchy and set of objectives rooted in memory. Most importantly of all, men follow the linear logic to the end, until the objective is achieved. It's hard to divert the attention of man once set on his course.

Insight into female methodology is also instructive, also has its beginnings in ancient, co-operative behaviour. Gathering food is a skill requiring an almost encyclopaedic, detailed view of possibilities. An impressive memory for detail and past circumstances is also important; where certain things grow, what season is best for which eggs and which fruit; when and where creatures might nest or burrow; which tell-tale signs are present or absent? A shared sense of the subtle changes of season, weather, terrain and vegetation is fundamental to the co-operative mood of women as well as men, and perhaps more precise.

A stable, open, conscious and unconscious evaluation of events, of current mood, shared associations and the dynamic of immediate or current circumstances is female, not some theoretical, pre-constructed cerebral goal. It is difficult to annunciate, categorise, relate. Much of its character is instinctive and, for want of a better description, hard to put into words, to order or describe a sequence. Sensory intuition is, by its very nature, hard to explain, justify, or rationalise. Thus, if it has no order, suitable explanation, such sensibilities are of no significance in a rational world, a world of linear logic.

Female thoughts and senses are debased, devalued, even ridiculed.

Yet, given their longevity, proven substance and value in long-term survival, their effectiveness must be undoubted. We know this because these sensibilities have survived for hundreds of thousands of years. That they still have a significant and relevant place in human understanding is a cogent argument for their value and unique origin.

That these innate sensory gifts don't easily satisfy the criteria for logical evaluation is something in their favour rather than the reverse: it's simply that half the population is unable to grasp their meaning.

'You never know what you are going to find', 'Anything can turn up' 'You have to be ready for an opportunity, size it up straight away'.

An eye for some small variation – the shoot of a new growth, the wilt of a stem, the bird droppings that might point to a nest, the scraping from a burrow, the bloom on berries about to become edible – all these details add to a woman's instinct. She has built a repertoire of crucial survival aids, honed by passing millennia, into an enviable if mysterious personality.

That women were probably more likely to accumulate deeper insights into the secrets of natural phenomena, and their cyclical repetition, is demonstrated by their different viewpoint as seen is the most advanced societies. They have not lost any of their inherent skills or abilities, but male linear logic overrides them. This need stems from a patriarchal imperative; to organise societies, to cope with the demands of their ever-increasing scale and complexity.

The danger is that women will lose their innate abilities as sexual characteristics become blurred, as we further eradicate nature from our consideration in favour of abstract constructs and reduction. In this sense we conquered her long ago; women face an onslaught of patriarchal tyranny that threatens to overwhelm a large chunk of human nature.

Fortunately, there remains in the human strong, even passionate love of nature. Remember, it is our mental abilities that have evolved rapidly, become more precocious while leaving behind our physical make-up, our 'natural self'. 'In the wild' as the TV jargon has it, humans survive only a few short days without access to – constant support from - the complex artificial niche we have devised for ourselves. Nevertheless, we still respond to natural instinct, a reservoir of behaviour as profound as it is ancient.

Coming face-to-face to nature is exciting, but only for a while. We enjoy it most when we believe, in front of a glowing screen, we can glimpse it's intimacies and dangers. The raw truth can be revealing.

Here's an example; in the higher slopes of the Pyrenees, undomesticated horses live in freedom. They are not so wild that they have innate fear of man. Quite the reverse; they are audacious and will roam among the hiker's cars and steal   food from their picnics.

But on the high slopes this access makes the sexual behaviour of these half-wild horses easy to observe, ideal for the lazy naturalist.

A herd of horses has a few males and a large group of females. One young insistent male pesters the females. There are pheromones in the thin air. Among the females, one or more is in heat, but only one will be receptive.

The stallion struts among the females, full of youth and vigour. Head held high, muscular and proud, mane tossing. The females continue grazing, pushing off the male if he comes to close, turning adroitly away, nipping at his neck or flanks.

Still, there's something in the air. The young male begins to home in on the female apparently more receptive than her companions. This is decided by a process of elimination; the least defensive is the most inviting. His attentions evidently stimulate her and she urinates, releasing even more pheromones. The excitement is now reciprocal – there is little now that can affect the outcome.

The stallion's penis has in the previous encounters extended, rather uncertainly, but now becomes semi-rigid as he struggles to mount. The mare is urinating again, but thrashing with her tail, spreading the urine in a spray. The male mounts again and at last is able to clasp with his forelegs the body of the mare. Semen is already being produced.

It's not luck but the subtle, precisely evolved forms of both male and female genitalia, without which the horse would not have survived so successfully. It means that, with surprising and efficient brevity, the mare is impregnated.

What happens next is instructive for the human observer; very little happens. After his peregrinations and final, successful climactic the male is exhausted, but is bewildered, too. The passionate, even frantic, drive that characterised his previous wild enthusiasm has evaporated. A vacant look replaces the almost fierce insistence of attitude and expression that enabled him to arrive at this point – his sperm has set in train the forces of procreation. He is now superfluous. He appears to wonder what to do next. There is nothing to do next. The paradox is apparent.

For the mare, however, far-reaching changes in the biological complexity of the female have been set in train, lasting not just for the term of the pregnancy. Only when the offspring is weaned and has learned the basics of how to be a horse, rain or shine, winter or summer, can the 'female' role, that is, in the restricted view of the male, be resumed.

The human view of sexuality is paradoxical; that it is a joint affair, a shared experience, meaningful and rewarding. This is at odds with the physical reality of what sex is about. The complex human view derives from aeons of 'humanisation'. Our clans and tribes prospered through shared social activities once language and tools were well established within an artificial niche. Recall drives a backward- and forward-looking bias that does not concern the horse, or any other species on Earth.

The hunters and the gatherers were significantly advanced; perhaps half-domesticated dogs trailed their hunting parties. Yet, within an era as yet unknown, perhaps spanning the epoch when recall first imbued humans with consciousness, the nature of the physical world became more apparent, also the context of humans within it, reproduction, birth and death.

Evidence of the already well-developed awareness of humans comes from their cave paintings. From these it may be deduced that certain aspects of humans' relation with the rest of the natural world are apparent; humans associated beginnings of life with caves, holes, burrows and similar features.

It should be remarked that most 'cave art', including etchings, marks, hand and thumb prints, as well as elaborate paintings and drawings, are often found in the most remote and inaccessible places.

Often the action of underground watercourses provided access for humans into deep recesses. These tunnels eventually become too narrow to penetrate for humans, but not for the watercourses that formed them. It is apparent that humans associated water-formed tunnels with access to a world that eluded them, which suggested the original internal sources of life itself.

Humans would remember that, in Spring, the young of bears, wolves other carnivores, along with bats, birds and a wide variety of creatures emerged, as it appeared to them, from a dark and inaccessible underworld. Human young emerged from the swollen womb, a mysterious hidden world that was also uniquely female.

Fresh water in the form of springs, streams and even sizable rivers, emerged fully-formed from clefts in the rock, bubbling from small apertures in the earth. Plants pushed upwards from the soil. Trees had their beginnings in small green shoots emerging from below ground.

Ground dwelling creatures such as insects, snakes, larvae, grubs, worms – all emerged from what would all too easily been seen as 'Mother' Earth. Right up to the time of Leonardo da Vinci, the Earth was seen as a living whole, an entire womb of life.

It is in this context that an inevitable separation of the human sexes might have formed; as the male completes his role, a sexual drive driven by long-evolved mechanisms fuelled by hormones and pheromones, a separate female psyche also evolves. Hormones and pheromones, but also a vitally more complex personal evolution that takes over where the male leaves off, his job done.

It is the female that is the carrier of the life process. Her early association with the mystery of the Earth and its natural forces must have been the beginning of a long association, still with us today (in the guise of the Catholic Virgin Mary, for example) as an entity uniquely different from the male. Across the species this fundamental difference separates man from the more vital processes of life. His mission is death.

This may seem melodramatic, but history and nature both reveal an essential truth; since the advent of recall, tools have encourage specialisation, as we have seen.

Gatherers, it should be noted, don't kill. It's closer to the truth that they garner. Females sensibilities and skills would enable them to recognise the value of cropping, not uprooting, of gathering not destroying. The female instinctively knows, would have known, that taking one or two eggs from a healthy clutch means another harvest next year.

These simple sensibilities are those that have guided females through long, hard centuries.

The male is intent upon the exercise of the hard-won linear logic that provides high-value protein. Recall in the male is a vigorous and powerful driver; survival becomes focussed into a single idea; the encircling of a prey by predators working to a pre-determined plan. Tools envisioned for a specific purpose outwit the wariest prey.

Early tools, we can see, mimic the 'weaponry' of much larger creatures. Flints become the lethal tiger's 'teeth' of men, more of them, too – longer, sharper, more deadly. Even the skins go to make shoes and leggings, jackets and (bound in strips) gloves and arm protection. Bones become spear-heads, antlers the primitive picks for unearthing flints. Strips of hide become slings, fur becomes protection from even the coldest weather. Wood becomes a primitive cart, sharpened poles a lethal defensive or offensive ring.

The prize is the biggest beast the hunters can butcher and carry off between them. The reward is the feast and the silent certainty that the puny figure of man is now the world's top predator. The psyche of the male and the female are already diverging.

CHAPTER 4

Woman as Goddess

It was at Naledi, 25 Kms from Praetoria in South Africa, that another piece of the puzzle emerged. The wrong turn on a research mission that led to the site where Lucy had been found many years earlier. The stone tools dating from 3.3 million years ago had their age confirmed. Over millennia – the Earth's magnetic field periodically reverses. North becomes South and, later, the field switches back. This provides a simple 'signature' that identifies deposits from a particular epoch. Earlier deposits carry a different (opposite) signature to later overlying layers, so a time-chart is written into compressed clays or sands that form rocks that can later be read like a calendar; a kind of layer-cake of time.

This recent (2015) addition of the earliest examples so far of manifest tools is of crucial importance. It's an indication of recall happening in the human brain. It takes us closer to the point at which humans became human, effectively leaving the animal kingdom behind.

It's also the first indication of how we began to distinguish between order and chaos. This may seem an abstruse point, but this ability defines humans. Order and chaos do not exist in nature – rather it is the recall ability that creates the distinction. Order is the illusion created out of recall; the process of 'ordering' events as they occur. Only in recalling the sequence of events does order 'appear' or 'make itself felt' in humans. To all other creatures the past is irrelevant and the future non-existent.

Let's see how this works. Our first 'garnerer' or gatherer, probably a woman, used the recall function to sort out her finds. She created 'order' because her enriched mental ability permitted 'discrimination'. Animal behaviour consists in simply browsing on plants or roots. Our *Femina sapiens,* browsing her environment, sees a hierarchy suggested by recall; which items would be better left for a later time, even a year. Location of plants (useful for fish traps or baskets) are stored away, because they can be recalled later from their place in a mental 'store'. It is the mental 'store' that provides order; a mental map of events, terrain and objects, their position in time and space.

With a new, enhanced mental repertoire, our lady gatherer is in a different class from her animal competitors. An animal's behaviour is less ordered, driven by hunger and instinct. Human behaviour is now more organised, ordered, determined and conscious. Humans now consciously affect the flow of events, shaping them to their advantage

The advantages are legion; by gathering and not eating on the hoof, our band is less exposed to risk . Eating a take-away saves time. Collecting armfuls of takeaways means the band can be safe at home for longer periods, share out food, perhaps digest more efficiently. It means a section of the group can do other stuff; making tools perhaps, weaving baskets, sharpening twigs or spears.

The key effect of recall is the mental ability to consciously scroll through past events and imagine future ones. This is how tools are 'thought up', how campsites are chosen, how hunts are planned, how foraging is carried out.

So recall creates order out of an essentially chaotic world. Our sense of order is by now so deep-rooted that it's hard for us to see the world as disordered, but actually chaos reigns. No two snow-flakes are the same, no leaf is the same as any other, nothing is duplicated, ever.

Of course, science, mathematics, engineering, logic and thought wholly depend upon a supposition that order exists. Our deep-rooted distaste of chaos produces a permanent mind-set that insists that order is a key ingredient in nature and drives the world. This, in reality (in nature as it were) is an entirely false supposition. Recall creates the illusion of continuity. In reality there is no 'time', place or object that is not a jumble of events and sensations that are impossible to revisit. The plumage of the peacock was not planned. Our brains, embedded in a world of order (born of recall) find it hard to accept that the peacock is the result of a series of timely accidents that allow the creature better to exploit a particular niche in a chaotic world.

Which brings us back to the world's earliest tools; they are the earliest evidence of the onset of recall. Though the long-lost wooden implements, of which all trace has disappeared, predate these by perhaps half a million years, perhaps much longer. Human consciousness may be 5 or 8 million years old. Hence, my claim that humans should be defined by evidence of tools, not by skull fragments, and that women should share equally in the undoubted significance accorded to the appearance on Earth of this unique species.

Over those long aeons of time it was the illusion of 'time past' and 'time future' that helped us change the environment to suit us, rather than us slowly changing to suit the environment. This crucial evolutionary advantage is only available through recall. It is recall that signals the importance of ancient stone tools via similar objects available to us today. We can identify easily those objects in the world that, like the time signature of polarity reversal, carry the evidence of human intervention.

In recent deep cave searches, prehistoric paintings have been found in such awkward recesses that only female archaeologists, with slender figures, have been able to wriggle through to reach them safely (and return!). In the caves of Mas d'Azil in the Ariege region of Southern France researchers found a female skull facing an engraving on the cave wall of a bison. Around her head a circle of flowers had been placed. Their dried petals were still there 50,000 years later.

Here we begin to see the first traces of ritual connected directly with females. Were the first artists female? Were they the priestesses who brought bounty to the clan by their intervention? Did their artistry and decoration embody the wishes and hopes of the tribe?

Perhaps ancient humans associated the underworld with mystery and thus femininity. The emergence of new life from the apertures of an unseen, unknown place equates directly (in the still-emerging consciousness of pre-history) with the mysterious behaviour and physical nature of the female; unaccountable bleeding on a regular basis, the protracted pregnancy, milk seeping from her breasts. The direct parallels with animal life, yet with powerful and creative consciousness too, suggest a being that is something more than just animal, more than merely human.

It is clear that *Femina sapiens* played a role denied her by modern interpretations of palaeoanthropology; she may have been quite probably the first tool-maker, quite likely, too, she was the first artist. Almost certainly the association of her with fundamental mysteries of nature imbued her with a significance that, to the male, was mysterious, not a little daunting, not a little scary.

The aggressive, projectile essence of ejaculation and use of hunting tools combine in a way that is also uniquely male, but also not animal. Modern males refer to their 'tool'. Men today exploit to the full the energy and power of the gun, perhaps echoing the 'action at a distance', the thrill of the primitive hunt.

This divergence of personality between male and female is so ancient that it is an embedded characteristic of the distinct psyches of male and female. Unlike animals, we share the recall mutation but our psyches have evolved along different routes through long-exploited divergent skills, coalescing into distinct character profiles, distinct sexual psyches. These different trajectories over such long sweeps of time defy change, despite our many attempts; despite subjugation, despite abuse, despite discrimination, the female personality, through its complexity and mystery, frightens the male. He reacts accordingly.

 The long and profound influence of prehistory tended in later epochs to enshrine the unexplained powers accorded to women in symbolism and custom. A kind of cultural shorthand. That she became a goddess is to be expected. So too is the calculated, cruel and persistent reaction of the male – the next conundrum we have to confront.

Many cultures worshipped or, at least, enshrined in deified form, an exaggerated personification of the female. The earliest known representation is a small, rotund figure with prominent breasts and stomach and only vaguely shaped head, legs and arms. Clearly, crucial characteristics held centre stage, equivalent to those qualities that preoccupy women today who seek the magic ingredients that go to make up 'the perfect woman'.

"The floor of the court had not been doomed to sterility by a stone pavement, but on the contrary, it burst with fertility, as behoves Aphrodite: fruit trees with verdant foliage rose to prodigious heights, their limbs weaving a lofty vault. The myrtle, beloved by the goddess, reached up its berry-laden branches no less than the other trees which so gracefully stretched out. Ivies entwine themselves lovingly around each of these trees. Heavy clusters of grapes hang from the gnarled vines: indeed, Aphrodite is only more attractive when united with Bacchus; their pleasures are sweeter for being mixed together. Apart, they have less spice. Under the welcome shade of the boughs, comfortable beds await the celebrants— actually the better people of the town only rarely frequent these green halls, but the common crowds jostle there on festive days, to yield publicly to the joys of love." (Pseudo-Lucian, Erotes) .
Christine Mitchell Havelock 'Aphrodite of Knidos'.University of Michigan Press 1995.

Fecundity, as it occurs in nature is no longer top priority. But, while the shape has changed, the essence of womanhood persists; the guile, magic, mystery, allure, some unreachable, inexplicable charm and power. Of course, men and women are programmed to recognise the ingredients and respond accordingly. Like horses, hormones and pheromones also play their part.

Nothing that simple can explain, however, the complex history of woman as goddess in the classical and later epochs. One archetypal female is Aphrodite, a particular favourite. There are many representations, derived from a famous sculpture by the Greek master Praxiteles, now lost.

Just to demonstrate the mysterious ambiguities of the archetypal woman, the sculptor's model was a courtesan, Phrygne by name. Courtesans were not accepted by female society in ancient Greece. But philosophers, poets, Statesmen, artists and sculptors of the highest repute were their companions. They were admired, loved, celebrated in poetry and literature, sculpted on this rare occasion and in this way elevated to heights and historic memory in a manner not accorded to any senator's wife or daughter; a goddess of unparalleled beauty, fully formed at birth.

Aphrodite of the legend was born in the sea spume when Cronus severed the genitals of Uranus and threw them into the waves. Many legends attend her, and the myth of her origin embodies a sense of fear and threat – what may be every man's nightmare is symbolically the prime ingredient in the creation of the most beautiful woman in the world in the 4th Century BC.

Homer even accorded her the status of 'the Earth Mother' and identified her with the Olympian pantheon of gods. The legends surrounding, not just her, but even Praxiteles' statue carved for her, are endless. Aphrodite became a tourist attraction

The statue became famous for its beauty, meant to be appreciated from every angle, and for being the first life-size representation of the nude female form. It depicted the goddess Aphrodite as she prepared for the ritual bath that restored her purity (not virginity), discarding her drapery in her left hand, while modestly shielding herself with her right hand.

From this we can judge the impact and influence of even a mythical beauty. Phrygne herself became almost as famous. But even though Aphrodite's admirers travelled to a distant island to view her statue, male contemporaries continued their insistence that females were markedly different qualitatively; men engendered the high ideals of a now civilized form of human, thus even their gods were become 'enshrined' in a hierarchy that has characterised human civilisation ever since. But yet, mankind could never rid culture – any culture, at any time – of the an innate fear, suspicion, a certainty lurking somewhere in the unconscious, that the female was the true embodiment of life, and the male, death.

So Greek men made war, using the very implements their prehistoric male counterparts employed in the primitive hunt. At heart, men were savages and still are. Still, Aphrodite strives for acceptance, even though her status as goddess wanes. Female mysteries are scattered by male attitude, the leaden implacability of linear logic rules triumphantly. Modern reason trumps magic. Only poor copies of Praxiteles remain; all that is left of archetypal woman, shuttered away in lonely niches in austere museums.

Further testimony for the continuing divergence of the male and female sexual personalities, their sexual psyche, resides in the origins of beliefs emerging as patriarchal religions. This, of course, is the male preserve, but probably born of the fear and mystery of early women as early men struggled with the dawning consciousness that women were inexplicable.

The nascent interpretation of sexual and social tension (felt by the male) turned toward demonising the female; a dubious but universal description from early history is that females really were 'the root of all evil'.

The story of Adam and Eve, the bible – Old and New Testaments - are inventories of how to establish and preserve a Patriarchal social structure; God is male, thundering imprecations at 'subjects'. Females are treacherous and deceitful, not to be trusted, only to be governed. Notice that the tensions of religious belief only begin to be resolved or softened, when (in the Christian religion) a male child, saviour of the World is somehow, contrary to any possible example from the natural world, born to a virgin mother. Of course, she could have borne a female child, but the female has already been relegated to the position of second-class citizen. So the male comes to symbolise the perpetuity of the male hierarchy, to be enshrined in death, so there can be no succession, ever. In other words, the male hierarchy must rule in perpetuity. Immortalized (but neutralised) the female and mother is, though with the palliative of goddess status, consigned to history, to art, to culture, to literature, to poetry; the 'second string' of cultures and their new materialism.

Thus, it is that females today continue, generation after generation, with no family name, no identity, no sense of place in the forward march of male history, condemned to the anonymous lineage of the 'captive wife'.

We have to return to the recall paradigm to unravel the reasoning, unconscious and primitive as it is in its origins and continuity. Though it is usually a verb, 'to recall', here 'recall' is used as a noun – there is no other means, except to use term 'episodic memory', coined by Endel Tulving. This term is unsatisfactory and, in my view, misleading, as it fails to convey the process and its permanent presence as an underlying function of consciousness itself.

The most important evolutionary pathway leading to the predominance of humans is therefore a mental one, with recall acting as the springboard for extraordinarily rapid human development. From our hominid ancestors, humans came to occupy a 'mental niche', the one we inhabit today. So, recall underlies all 'ordered' mental activity, be it rational thought, language, the concept of number and logic and, consequently, all science.

If episodic memory is a separate entity—anatomical, homological, physiological, or in some as yet unknown fashion—in the normal brain, it is possible for damage to the brain to occur in such a way that it is deleteriously affected while other kinds of memory are not, or are less affected... (..episodic memory emerged as an "embellishment" of the semantic memory system. Ibid.). Tulving. Annu. Rev. Psychol. 2002.53:1-25

Recall produces a powerful illusion. In this way people in more recent periods of history could predict the seasonal cycles. They gained the status as priests; information became associated with specialists. It still is.

As Peter Ackroyd writes of the 17th century, "in truth it was century in which science became the new form of religion". Now 'new' men such as Isaac Newton dealt in matters of 'unassailable dogma', only it was science, not religion. And it is this phrase above all others that characterises how '*Homo sapiens*' thinks, how he is set apart from any notion of female intuition or flexibility of intellect. Dogma, linear logic and certainty rule the male sense of purpose.

(It should be noted here that '*Homo*' is derived from the Latin for earth or soil, but after such long association with male orientated science it is now firmly associated with 'Man', hence my choice of *Femina sapiens* from Latin: 'wise woman')

We can trace the origins of this mental approach back to the highly successful recall process that produced tools and weapons, as well as the strategy for their most effective use. Males are hostage to this view of the world and it is with great difficulty (as many women will testify) that they can be diverted or distracted. The 'modus operandi' of males is to asses possibilities, read the territory, come up with a plan, make sure he has the right 'tools for the job', not be diverted from the goal and pursue it until it is successfully accomplished.

At the wheel of a car, and lost, only with a great effort of will can a man ask for help. In the supermarket, the trolley becomes the tool that drives their actions and the checkout the goal that must be obtained in (to them) the most effective manner.

This is the linear logic learned in prehistory and relying of the potent force of recall and the tools and mechanisms so derived. It is still with us today, as lethal or productive now as it was then. The male is captive to linear logic and may be diverted from it only with great difficulty.

Here we come to another crucial part of this analysis; science, too, is firmly rooted in linear logic. The 'linear' element is an addition that emphasises both the procedural character and structured method that is directly dependent upon recall. The 'terms' of formal logic are the accepted definitions employed in reasoning.

There is no need to revert again to the dusty origins of the human consciousness.

Between our first ancestors and Isaac Newton, populations of males have existed with an insatiable desire to continue along the path of linear logic. Recall dislodged them precipitately from animal-like, dogged instincts honed to deal with the chaos of nature. We return, again and again, to the successful formula. In its 'pure' form, linear logic equates with 'absolute certainty': given the abstract nature of mathematics and self-consistency, you can 'prove it'. You can prove anything. Rationally, everything 'falls into place'; is, must be 'true'. Humans – particularly men, particularly scientific men – are hostage to this particularly advantageous piece of fortune.

Chaos can be said to characterise the natural world and the universe. Order only exists in the human mind. 'Certainty' or the illusion of certainty is the inevitable consequence of recall. It seeks out the suggested connection between events. It enables us to categorise them into self-consistent ideas. With recall, humans automatically group living entities, ideas, objects and phenomena, assembling these into familiar ingredients in day-to-day living. This is the basis for thought, language and, ultimately, science.

In his notebooks, preserved in the Darwin Archive at Cambridge University, notebook 'M' bears, among many, one crucial entry reproduced here, hitherto overlooked by evolutionary scientists.

Darwin, speaking here of the 'comparison with past ideas' (author's emphasis) implicitly acknowledges that only with recall can consciousness exist. He failed to realise also that consciousness is unique to humans; no creature in his exhaustive studies exhibited the ability to record thoughts as he so famously did.

"The mind thinks with extraordinary rapidity — We may conclude that neither number, vividness, rapidity, novelty of separate ideas cause fatigue to the mind, — it is solely the comparison, with past ideas, which makes consciousness — & which tells one of reality"

Darwin, Charles. Notebook 'M'. [103]. Cambridge University Darwin Archive.

He failed also to realise an important part of the recall paradigm, had he pursued its implications to human consciousness: *the act of writing makes thought permanent*, an idea perhaps too subtle to penetrate the mind of the scientist driven by linear logic.

In November 2015 the world learned that one of four surviving white Rhinos died in San Diego zoo. To all intents and purposes, these creatures are now extinct. Humans have hastened their demise. Few, if any, have been eaten. The contrast between animal behaviour and instincts is in stark and unnatural contrast to that of humans. 'Unnatural' because, while there are 8 to ten million other species inhabiting the Earth's biosphere, only one has the mental ability of man, an ability the product of a mutation. It's time to reassess the true nature of the human species and our impact upon the rest.

Darwin would resist the idea that in some way humans might be a flawed species, and he's not the only one. Even as he struggled with the completion and publication of 'On the Origin of Species' Darwin sought to reassure Emma, his wife (a devout Christian). Humans (he believed) were indeed the final and most powerful expression of natural selection; that species evolved as a tree grows, upwards, ever higher and that the pinnacle of perfection and, above all, most natural was the white, human male.

We can forgive Darwin his arrogance. It was a characteristic of the time, just as retrospective moral judgements are a favourite of ours. We can add that Darwin was unaware of the effects of types of radiation now known to cause mutation, resulting in physical changes (or in the case of humans, perhaps changes to cerebral architecture) that environmental pressures may favour.

In the case of humans, just such a mutation allowed us to sidestep the slow processional change that Darwin so well described. He failed to realise that while the 'small changes' and 'slow degrees' of evolution applied to all other species, something else occurred to produce humans. The irony is that Darwin, in an obscure and overlooked sentence in notebook 'M', hit upon the answer but did not realise its significance, nor could he explain the mechanism.

Darwin knew nothing of DNA or the spiral molecule that turned out to be the human genome. Retrospective science is just as unprofitable as retrospective morality; great discoveries of the past look tame and tiny from a modern perspective, yet Charles Darwin is a hero of the scientific age and of natural science.

CHAPTER 5

Mrs Darwin

Emma Darwin might be said to have been famous in her own right, but only via the name of another, by being hostage to male fortune; Josiah Wedgwood was her grandfather.

The Wedgwood and the Darwin families had a long association, cousins marrying cousins in the preceding generations. If Darwin had known what modern genetic science knows, or had Emma, Erasmus Darwin, or Josiah Wedgewood or Charles himself, then they would have realised their marriage to be ill advised. Much of the tragedy of the family's lives might have been avoided.

We can point to Emma Darwin as a heroine of the age, too. Quiet, reserved and highly respectable, Emma encouraged and helped Charles in his work, supported him in society as far as she could. She was 'of the time', for although she resisted bitterly his theories that challenged the role of the 'Creator', was entirely loyal to her husband. She kept objections to their private conversations and letters but also saw that the beliefs they had both once shared were also a source of his anguish and uncertainty.

Darwin, then, was also a male not be deflected from his purpose. Emma was the woman who would and could suffer stoically everything the world could do to shake her faith. She was unshakable, even though her children suffered too. The closeness of their DNA resulted in a number of children, some healthy, but others either handicapped or doomed to die; one son dead having failed to develop (as far as we can tell) even a brain. In the light of this, what was the purpose or the significance of untried theory, one that flew in the face of everything Emma believed in?

Emma's faith never wavered; perhaps because she identified the 'punishment' suffered by the family and her children as divine in origin? Charles' suffering led him to seek physical cures for his constant stomach upsets and a morbid fear of appearing in public. Presenting papers at the Royal Society were a considerable trial.

Darwin knew his theories contradicted the widely-held views of Christianity. He had intended to study divinity at Cambridge before a fascination with molluscs diverted his attention to Natural History. It was as a young adventurer, freed from the prospect of a life as a

Churchman, that the position as Naturalist aboard the Beagle (on its surveying mission to the South Atlantic) liberated his true intellectual capabilities.

In sense, he was already betrothed to Emma. The Wedgwoods and the Darwins, like many established families of the period, through custom and prudence, chose their spouses from a select but limited number of candidates. Darwin was not about to become a swashbuckling globetrotter, marry a native girl and live happily forever on a South Sea fantasy island. He was already a man, a naturalist trapped in linear logic, even before his marriage.

Emma, too, was already a captive to her sex. Custom, social pressure, geography, all decreed there was no other man for her. There was no choice. Better, though, if Charles had stayed the course of Divine Instruction and she had married a parson.

The huge fossil bones of dinosaurs Darwin sent back to England via passing ships were fascinating the public (he first became famous in London as 'the Fossil Man') but mainly because they were believed to be remains of creatures drowned in the Biblical flood.

It was in this climate of new science, driven by such intellectual engines as the 'Lunar Society' (with luminaries Josiah Wedgwood and Erasmus Darwin among its members) that the intellectual renaissance of late eighteenth and early nineteenth centuries, collided head-on with the deep-rooted history of religious faith.

History hardly mentions the women, but the social hierarchy of the day can be seen, by the doubtful standards of today, as rigid, un-yielding, and claustrophobic. Yet this period was, in terms of human progress, one of unparalleled advancement.

Darwin's energies, we can see, demonstrate the established pattern of linear logic driving that advance. Emma, like many women of her time and generations of predecessors, plays a subsidiary role, more reticent than any women of the present. Nevertheless, the epoch exploited the ancient female skills; the very processes that Wedgewood introduced into industry were ideally suited to the innate female abilities, their skills and character. Early mass production, though directed by men, was also the preserve of women.

The attention to detail, the eye for small variation . Like the garners of millions of years before;

"...these are fit, sturdy women with keen, accustomed eyes. They have hair that is short or tied back out of the way. It's immediately apparent they have an eye for detail, a demeanour suggesting deep concentration, experience, knowledge; they know what they are looking for."

From an earlier chapter, a vignette of the female describes the characteristics that the 'new manufactory' found to be so effective and profitable. The daughters and wives of independent countrymen and women, paradoxically, became the 'slaves' to the linear logic of mass production. Migrating to the new industrial towns, women displayed a natural aptitude for industrial methods; from Wedgwood's pots to acres of woven cloth, endless skeins of wool, fabrics, canvas, linen, cotton, and cutlery, *Femina sapiens* drove on the rapid success of human-kind.

Female dexterity and stamina brought herring by the ship-load to market, gutted and packed by women following the seasons and the fishing fleet along the coastline. The skills of women were thus resurrected, refined and preserved. The impetus and psychology that drives the men is also the same as in an ancient landscape. The early industrial revolution was also the unnoticed evolution of an artificial niche inhabited by humans. Industry demanded an increasing number, variety and complexity of manifest tools.

Emma's letters and diaries chart the course –in outline, at least - of a domestic life through this turning point in human history; first, we come to terms with ourselves. Second, the implicit comfort suggested by the presence of an Almighty Overseer is defunct.

So began an epoch of human arrogance unimaginable by the most ambitious dictators of the ancient world; the concept of mass production itself was chosen as a model by a new class of political thinkers. Liberated by the modern, rational view of nature, and the irrelevance of any divine influence, workers came to be viewed by that other new class of liberated intellectuals as mere 'factors of production'. Viewed as commodities, or as diminutive machines, new methods turned work into repetitive actions, reduced to operations easily measured and controlled, even by unskilled labour. Not just factory owners benefited from this new niche structure. Political and social engineers, philosophers and political experimenters viewed the new populations as grist to their intellectual mills. Large numbers of people, now deprived of any social status – equalised, we may call it - were cynically manipulated and unconsciously recruited to a new religion; the bald logic of social politics and industry. The industrial process and the fate of their 'mass-produced' operatives still fuels the exploitative political rhetoric of 'The Labour Party'.

In this way, small bands of humans, migrating from their villages, country plots and families - migrating through aeons of history - entered into the maelstrom, an era with an evolutionary outcome more turbulent and precipitate than in any previous epoch.

It is a truism that, today, the children of the industrial age are barely able to identify the origin of milk, bread, meat, beer, honey, wool, steel or leather.

Any benefits from the modern obsession with excessive production may be debateable, but the process of industrialisation separated humans more decisively from the more familiar niche of country, weather, plants and animals. The city became the new niche. The impetus must stem from an ancient driving force; the need for security, to gather for tomorrows need, to guard a meagre surplus.

Then came industrial warfare, ushered in by the mere fact that even 'action at a distance' can be mechanised.

Tribal battle and hunting takes on a new dimension. Ordinary men, displaced from any real involvement in the challenges of the natural world, relished the chance (at first) to test their metal and to extend their influence across new territory.

Ordinary women welcomed the opportunity to see their chosen mate in a more basic, appealing form, in which roles were clear and unambiguous; men reverting to their original personas; aggressive explorers, ingenious and lethal hunters, dedicated protectors.

The appearance of linear logic in warfare is both to be expected and to lay bare primitive origins; instinctive behaviour dominates when all pretences of a complex society are stripped away. The appearance of sophisticated industrial weaponry does not disguise the primitive nature of its 'action at a distance' function; the spear may be 'longer', the target farther off, yet the principle is perhaps three million years old, as is the logic.

Except the industrial scale of the 1st and 2nd world wars introduced an entirely new and terrifying dimension to 'the hunt'; collateral damage that was 'off the scale'. Women were not 'safe at home'. No longer could the old and infirm be protected and cared for, the tribe could not exist amid the sheer enormity of the destruction. Here the leverage effect of recall is laid bare – simple, primitive ideas 'scaled-up' to unbelievable and irrational proportions, gratifying a visceral, primitive compulsion so ancient that it is impossible to eradicate.

The quintessential expression of linear logic, unleashed by those who wished to explore its limits, came close to exterminating the very people who were most thrilled by its power – namely men; primitive, visceral hunter-killers.

Part II

<u>THE PRESENT</u>

CHAPTER 6

<u>Chaos and Order: Male Fundamentals</u>

As we have previously seen, male identity centres on an obvious piece of anatomy, visible to the naked eye, as it were, whereas visible female sexual characteristics are largely (or not) the breasts. She may be hirsute but this is adjustable. Men, have mostly hairy faces, sometimes chests, arms and legs. They are more muscled, generally bigger-boned. Intellectually we are pretty much equal.

Mentally, and from the point of view of the psyche, sexual drive is persistent and, for most of adult life, more or less permanent, given the right stimulus. Sexual activity can be more frequent and often of surprisingly short duration. The mental modifier of sexual behaviour can be forceful or absent.

These are broad descriptions, not definitions, and are imprecise to the clinician or the dedicated enquirer. One aspect of male psyche is a strongly defining characteristic.

Recall acts on the male psyche over evolutionary timescales to favour behaviour of a special kind, special in that it is distinct from that of the female. To describe it we must revisit the 'recall paradigm'. Recall as we have seen is not a 'natural' phenomenon in the sense that it is present (as far as we can judge) in any other species.

Perfectly successful species unchanged for millennia have no need of recall. They do not examine past events to see how to change the world so that it suits them better. They are perfectly adapted to exploit the niche that scorpions occupy. Recall changes memory – the store of stimuli that the scorpion lives by, the smell, the objects by which it navigates its niche, that are built-in, 'hard-wired'. As these objects are presented or present themselves to the scorpion, reactions are produced which favour or do not favour a positive outcome to support the continued existence of the scorpion. Niche and insect combine to construct a continuum, the scorpion's life. The scorpion doesn't have to think or plan.

Whereas there are 8 or 10 million species of plants and animals, one can say with certainty that none of them think. This may seem an extravagant claim, many will say 'my cat' or 'my dog' thinks, that elephants, horses and dolphins 'think'. However, thinking is not musing or dreaming, the most we can say of what happens in an animal's head.

Animals birds or insects, even apes, do not mentally review the relative importance, the sequential order of events in a way that might lead to something entirely new and then go on to produce it. All this can be accomplished independently of any external stimulus.

This is what we mean by thinking. Animals of whatever kind do not re-order imaginatively events, sequences or structure. They do not scroll through time past 'in their heads'

HOW TO 'BY-PASS' EVOLUTION

Many behaviourists have exhaustively studied many animals, higher order animals included, and continue to test them, which perhaps says more about humans than animals.

They have discovered intelligence – of the scrub jay, parrot, dolphin or ape kind, but not human intelligence. This is the ability to think in the abstract, to recall events from the past voluntarily, to classify or re-order them at will, to determine future out-comes.

Nature slowly evolves animals to better suit a particular natural niche.

Humans use their intelligence to create their own niche, adjusting nature so that it suits them better, by-passing evolution.

to pick out a sequence or series of past events and from this experience then assemble (still in the abstract) some ordered outcome or something of novel use. They do not employ imagination in order to devise a future plan, to order circumstances in a way that is solely the result of their mental efforts. We do this all the time.

The study of animal behaviour might be the result of our sense of the extraordinary disconnect between nature with its 10 million species, and us. The recall mutation helps to explain it; we are of a different order, a product of evolution evolving.

In the human, the recall gene (there must be a gene somewhere expressing for recall) has become specialised through long use from an early time. It's important to note that, while modern humans have been around for less than fifty thousand years, Lucy and her clan spent three thousand, five hundred times *as long* evolving towards the modern human and they had already cracked tool-making.

Recall set humans apart but it also set the male apart from the female, forming a different psyche, a different mind-set. Recall creates linear logic all by itself. We all instinctively know how it works, it is so familiar that we don't have to be scientists to discover its compelling power. A difficulty calls to mind the sequence of events that led up to it. Mentally sifting through similar experiences we, and our primitive human, use recall to compare like with like; a previous successful outcome with the one that didn't turn out so well. Perhaps it was an early bow and arrow with the string too slack, the bow too short. We, or in the past, he or she, are able to try another version. We can test it without even the presence of prey, or the problem, present. We can do it " all in our heads". This is where the true definition of humanity is found, in what I call the Recall Paradigm.

By slow degrees, improved hunting success (the preserve of the primitive human male) provides more food. More food means more time to plan because hunting becomes more effective, needing less time. These cumulative benefits encourage more planning, further increasing success. No wonder, after 3,500,000 years, linear logic is the trusted modus operandi of the human male.

Recall of course accelerated the success of garnering and gathering: a stick cut off at the fork can do as a simple 'plough'. Such a light and simple device can fashion a rut in soft soil, enough to replace surplus seeds and provide for the future. Such a strategy is more probably female in origin. Perhaps women invented the plough and, thereby, agriculture while men were out hunting high-value protein.

To emphasise further the effectiveness of the strategy so characteristically male, it persisted until what me might call the modern era, which, if measured by the use of say, the bow and arrow lasted until the introduction of the cannon at the Battle of Crecy. The simple wooden 'plough' lasted until well beyond the iron age.

Firearms are, have always been, a preoccupation of the male. His psyche has a special place also for guns, bats, balls, cars, aeroplanes, missiles, radar, telephones, wireless, anything, in fact, that can be characterised as effecting 'action at a distance'. This is the prime purpose of hunting technique, to use a device that kills the game without endangering the hunter.

The shape and projectile action of any kind of similar 'tool' is seminally familiar to the male. They are part of his sexual psyche.

That he adopted early on a powerfully effective piece of original thinking , the projectile action and its subsequent powerful role as 'spreading the influence' of the operator - is evinced from its use very early on. The earliest 'action at a distance' methods would not survive the intervening epochs. Such simple devices as pointed wooden poles and hunting weapons of wood like the bow, the slingshot or snares, easily decay and lost forever. The snare is purposely included in the 'action at a distance' category; the snare is set, it works (hopefully) while the hunter is elsewhere, the prey retrieved later.

Spear-throwers represented an advance, around 40,000 years ago, and there are beautifully crafted examples at Mas d'Azil. These were short-notched sticks that effectively increased the velocity of the weapon; a sling shot effect. The stylised carvings of animals no doubt increased their effectiveness.

So the male psyche is consistent, if nothing else, and, because linear logic is an ingredient of every event the male is likely to encounter, no wonder it is central to his character, no wonder it is so hard for the male to respond in any other way; linear logic works.

Linear logic has always been the impetus for new ways of doing things and the scientific method is the method whereby new discoveries occur. For those proven 'successes', the stored memories of collective humanity are carefully sifted. These tried and tested outcomes of science lead to new theoretical propositions. Proof begets new proofs. This methodology is the means that brings forward carefully reasoned propositions. Self-consistency in linear logic is the basis for science in theory and for technology in practice.

I mentioned there would be trouble ahead: if science is simply a product of recall and, as recall is the result of a mutation, is everything in science simply an illusion, based on a mental flaw? Is 'reality', in fact, 'chaos'; natural events simply 're-ordered' via a unique mental process to allow us humans to gain an advantage over all other species? The answer is probably 'yes'.

Certainly, for hundreds of generations of widely dispersed humans, the illusion of religion for example, drove the passions – many violent – of a vast disparity of cultures, often separated by continents and oceans. Religious belief determined the lives of millions and the deaths of millions more. The idea of a human-like mind 'controlling' events was then, and still is, the common denominator.

The power of religious belief to direct the actions of humankind across swathes of time and humanity is undoubted. Now the power of science fulfils that function. Today's high priests aspire to a similar role. The unshakeable belief, the promise of domination over all events in nature, the promise to conquer all enemies, satisfy all desires, solve all problems, is almost the sworn creed of the scientist. The unshakeable faith in the human intellect generates a quasi-religious fervour, which in some is frightening to behold.

The creed that drives this modern religion is linear logic. It is the basis for the widely-held belief that humans are actually, from our perspective, a kind of sublime, all-powerful deity. Many scientists see religion as a threat to their presumed supremacy, perhaps as subtle acceptance that they have usurped the role. Many are even vociferous deniers of 'faith' in any other form. Their ire is reminiscent of religious fervour: 'There shall be no other God but me'. As we have seen, the logical case sits upon some very uneven ground; there are fault lines below, there is no underpinning of solid evidence when we investigate human evolution.

The recall paradigm, even as a tentative thesis, undermines the notion that 'logic' exists as a pure, unadulterated, pristine form of reasoning that makes conclusions undeniably 'true' and therefore (in a religious sense as well as practical) undeniable. The history of science is replete with examples of failed concepts once 'undeniable'. The only undeniable fact is that science is arbitrary, that evolution is the final arbiter, the 'niche' the agent that shapes the life form.

The human mind alone produces science, mathematics and logic. The human mind is a product of nature, the chaotic collection of events and outcomes. All we do is survey the chaos and *select* from what we need in order to exploit fully the Earth's resources to our sole advantage. The recall mutation may have separated us from all other species and by a wide margin. It did not make us Gods. It is simply that we live in a 'mental niche', the product, like all evolutionary change, of a mutation, shaping the organism to be able better to compete for available resources, to survive.

Serious problems begin to emerge that are inherent within the god-like ambitions of science and technology, as we are now beginning to see. Reliance on empiricism is fraught with contradictions that are often 'swept under the carpet' to avoid any embarrassing logical challenge to the next great advance, development, theory, discovery.

The logic that underpins reasoning and science has always contained 'fault lines'. Hegel disagrees with Leibnitz. Friedrich Nietzsche was criticised by gangs of philosophers and logicians for declaring; "Logic came into existence from men's heads out of illogic, which must have been immense". And, in "The Joyful Science"; 'Logic, too, also rests on assumptions that do not correspond to anything in the real world".

Women, not uncommonly, are often characterised as being 'illogical'. The above telling quote from Nietzsche may shut the man up (it's always a man) and champions of logic are no longer in an unassailable position; its principle champions, theoretical and applied science and technology, stand condemned of accelerating 'to its logical conclusion' the rape of the world in favour of mankind.

It is unfortunate that males are guilty by implication of citing logic as justification for actions inherently counterproductive in the long term. They are hostage to a way thinking that is outmoded.

It is certainly the case that we can accomplish extraordinary feats of endeavour following through linear logic. Literally, we can move mountains. We seldom ask the question 'do we need to move the mountain?' The thrill of the task, the scale, the enormity, the danger, excites us, just as we suppressed fear in the thrill of the chase to surmount any danger in favour of high-value food. A fascination with the challenge itself is enough to prompt dangerous, dare-devil, physically violent, exciting, exhausting and 'rewarding' outcomes.

Sometimes we contrive conflict merely to re-enact the 'thrill of the chase'. Gang war, turf war, or just 'plain war', even a cursory assessment of male behaviour presents us with this disturbing conundrum; that humankind, usually headed up by striving males, has the strength, power and conceit to assume direction of human affairs, from the tribal to the global, at the expense of every other inhabitant of the Earth.

There is no limiting influence on man. Once humankind crossed the boundary of natural limiters, the confines of a physical niche *in nature,* there was no going back. Recall was responsible, it seems, for that crucial step, and there is no way back. Unless... (we'll return to 'unless' in a later chapter).

Chaos and order constitute the 'front line' in man's battle with nature. From earliest times to the present, it makes little difference whether the challenge ahead is reports of a woolly mammoth over the next hill, a prospective visit from mother-in-law (not necessarily the same thing), or that Mars glows temptingly at some improbable distance. The masculine approach is to 'stiffen the sinews', prepare for the challenge ahead.

With no challenge ahead, the male becomes disconsolate, feels robbed of his identity. The role of the male is not to accept the status quo, rather to meet head on the challenges of survival; for him, 'survival' *is* the challenge and has been since pre-history.

Many young men, having no opportunity to take on a challenge in a severely ordered society, may feel there is no role for them and descend into depression. The challenges of the natural world, with chaotic ingredients that need ordering, have in many cases ceased to exist or neutralised to the point that the male is unable to pursue any other path but to invent new ones, or even simply imagine them.

When Nietzsche writes, 'logic came into men's mind out of illogic', he is, unknowingly, making a direct reference to the evolutionary mismatch caused by the

103

recall mutation. When men refer to women disparagingly as 'incapable of logical thinking' they are unconsciously acknowledging the existence of a separate stream of human consciousness that may be better suited to understanding of, empathising with, the natural world. Chaotic structures frighten men; they are inaccessible to linear logic. For women, sensory intuition suggests that chaos is the natural and authentic condition of the universe. They can deal with it. Left to their own devices, nature's vagaries do not evoke fear, raw nature is in their blood.

So this other type of human has a pedigree as long and as noble as the male. The female doesn't have the restricted view imposed by the visceral need to follow 'linear logic'. Neither 'action at a distance', nor the need to construct an 'explanation' and thereby resolve uncertainty or mystery is of any concern. She has no need to out-face fear and danger through aggression, mental constructs, or a grand design that explains all.

CHAPTER 7

Sex and the Sociotype

Genetic make-up, psyche and insights gained from many aspects of life together add up to intuition. Where linear logic doesn't serve, intuition may often work better. Of course, from the masculine point of view, this definition counts as meaningless. In deference to male readers, therefore, it deserves amplification.

It may be that distributive intuition is a throwback to a primitive life, as primitive as spear-throwing or navigating the wilderness with a blazed trail.

It demands a highly developed set of senses, yet deduction doesn't help; it's too laborious and time-consuming. Just as men become impatient with a woman's supposed 'lack of logic', women fail to understand why a male can't grasp instantly the real nature of circumstances, appreciate the significance and vital importance of seemingly trivial instances, grasp the connections that are so (to them) blindingly obvious .

The term sensory 'intuition' is used to encompass the great variety of ingredients; nuances, hints, echoes, flavours, relevance, colours of a vast range of memories and stored reactions that constitute a 'repertoire' of intuitive skills driving interpretive behaviour that is anything but 'linear'. Rather, it is the absolute opposite.

Women frequently use the epithet 'multi-tasking' to describe their particular skills but this, I believe, is inadequate as well as inaccurate. A woman does not need to be 'tasking' while a complex set of events unfolds. Sensory intuition is the product, often unlooked-for, of a refined set of sensibilities, made up of unconscious processes, both physical and mental.

It is not the case that women have no grasp of logical behaviour. Many of the best 'legal brains' are women, perfectly capable of producing or following complex, formal and intricate reasoning processes. However, it is a 'truism' that when necessary (presumably not in court), women can take a sensory 'short-cut' unavailable to men. It is another truism that women are exasperated by the male's inability to follow them.

Partly, this is a result of an inability to explain the 'short cut' method. The problem is there is no 'method'.

Male linear logic even dominates the academic study of personality, psychology and socionics. Male theorists laid out the ground rules for both sociology and psychology. Even when female academics formulate theories within the social sciences they do so within a male intellectual framework.

Here's an example from socionics, the modern view of classifying personality types, now extensively employed in deciding, for example, the best members to recruit for a team manning a spacecraft, supervising a nuclear power station or a new marketing group.

First theorised by Lithuanian psychologist Aušra Augustinavičiūtė, the classification of human 'sociotypes' has led to a system allowing predictions of human behaviour according to 'types'. Types are characterised by definitions of their psyche or personality. Thus, an intuitive ethical extrovert (IEE) might have a social role as a Psychologist or Reporter, while an ESE, an ethical sensory extravert, is a 'Bonvivant' or Enthusiast.

Of course, while there is a good chance such classifications are sound to a degree, a logical conclusion might be reached but self–consistent only with a doubtful set of principles.

Individuals selected by these methods often have to take prescribed tests, also designed to confirm logic, standardisation and reproducibility. So selection is by 'forced choices', that is, only two choices are available to the subject; 'do you prefer sweet or bitter drinks?' 'which of these colours do you prefer; red, or blue'?. There are no other choices.

So a kind of shorthand, characterising just 16 different facets of humanity, was devised; Augustinavičiūtė must have believed only limited characteristics describe humans; Extrovert and Introvert, Logical and Ethical, Intuitive and Sensory. By mixing these varied 'functions' she arrived at the sixteen sociotypes . Thus, in her scheme, a Sensory, Ethical, Extrovert is characterised by a Napoleon or Caesar-like character, while an Intuitive, Logical, Introvert is more like Balzac. Ironically, though history is not lacking in prominent female personalities, she mentions not one in her lists of sociotypes.

As so often in science, we find only linear logic as the dubious guiding principles of thought, discussion, research and theory. Often, it constitutes an entire culture dominated mainly by men, or by characteristically male reasoning. 'Socionics' is an example. As male thought dominates science in contemporary affairs, so science is the dominant cultural force, mapping out our future, shaping the reasoning of female researchers and scholars, too.

Science fails, then, adequately to represent half of the human population in its attempt to find a logical method of describing behaviour or personality, or predicting with any degree of precision, what behaviour might be expected from them. Here the example, admittedly an extreme one, does nothing to dignify scientific method or allow any discussion, once the method is secure, which turns out to be its main purpose. These are the laws of science, after all.

Other areas of science pose the same problem; a kind of myopia in which arriving at a definition categorically rules out any other assessment, opinion or appraisal. Indeed, any approach not scientific is denigrated, dismissed, ruled out, disallowed. Even if thoroughly scientific, if the lead scientist is a female, attitudes count.

A notorious example is the 'discovery' of the double helix by Crick and Watson. The 'discoverer' was Rosalind Franklin. They excused their patronising attitude (and of effectively excluding her from the publication of the discovery) by citing the conventions of the time regarding gender in science.

The breathtaking arrogance engendered by the sanctity of certainty enshrouds all science. The close-knit coterie of the science establishments rests on an assumption that conclusions, based on self-consistent principles of reasoning are inevitably correct. Unfortunately, this very same method, which accepted as true all previous assumptions, are subsequently found to be inaccurate, or downright false.

The un-challengeable nature of modern scientific pronouncement mirrors the potency and force of an extreme religion.

Only science has 'meaning', in contemporary culture, everything else is seen as mere 'personal interpretation' and one can see the effect in painting, music, literature, architecture; there is a brutality beneath human affairs leading to empty mechanistic outcomes. Thus science, through self-imposed austerity and rigour, effectively removes 'the human' from 'the equation'.

Regarding scientific method, Nietzsche's conclusion is damning;

"..a mobile army of metaphors, metonyms and anthropomorphisms – in short metaphors which are worn out and without sensuous power; coins which have lost their picture and now matter only as metal, no longer as coins".

Now the human identity, both male and female, is being lost as a core social mediator; overshadowed by science to such an extent that many sense that a remote priesthood rules, the primitive individual re-emerges, with its primitive drives and appetites. No physical or moral restraints exist to prevent the deployment of an infinite array of powerful resources, hitherto available only to uniquely powerful individuals. Now every terrorist has a satellite 'phone. Cyber crime is a global scourge.

Human fascination with the cult of science has a lot to answer for.

Many women probably 'act out' the mannerisms characteristic of the male, adopting their behaviour and aggression, choosing this method of compliance to avoid confrontation which seems 'unwinnable'. This is the only route to power and influence, for the notion of female power – power that is feminine - is now untenable, seen as self-contradictory. Power is measured in a scale of male values.

Vital 'tasks' and 'roles' are fundamentally male, usually prescribed by the linear logic of their world, not seen by women as effective but accepted because they have 'no logic' to counter with. In the present world, sensory intuition is unlikely to survive the test of linear logic. Unlikely to win an argument today, sensory intuition was once likely to be a core survival skill. In some unique women, it was believed these gifts derived from a closer relation to nature than was possible for any man, no matter how powerful.

The reverence for ancient wisdom has gone. But, the evidence remains in surviving cultures and religions that female gifts were, and are, mysterious, powerful, magical. Like the Virgin Mary, women as queens, oracles or priestesses were hallowed, revered and worshipped.

Once, not long ago, a sculpture of a woman – courtesan or goddess – was a cultural treasure, capable of lifting a nation to new heights of influence and to global renown. Canova's Three Graces was such a work. The idea of creating such a work today is laughable. Once more, a patriarchal tyranny is the dominant force in human cultures.

The power of linear logic has robbed the world of sensory intuition. Yet the patriarchal triumphalism of this kind of reasoning is sadly hollow. Humankind clings to the notion that nature is a perpetual foe. In fact, evidence points to humans themselves as being 'their own worst enemy', devising solutions to non-existent problems. Small-scale occupation of parts of the biosphere provides an ideal living environment for humans.

The evidence points to the male sexual psyche as a direct impediment to continuing peaceful use of the Earth's resources. It likely that women intuitively hit upon practical solutions in devising the first wooden tools, woven fish traps, grubbing tools, vessels, or roughing out the shape of suitable skins for warmth and protection. Their instinct is for the sustainable. The male instinct is to arrive at a finality.

It is unimaginable that men would bother with the fine detail of a sustained, natural existence. The 'heavy lifting' of survival was down to them. Their instinct prompts solutions aimed at a single outcome. For women there is no problem, the 'outcome' is not necessary, only the continuity.

Here we can re-state the case from the male's dominant place in finding places to sleep and eat, to find water and navigate rough landscape or ward off predators. As we've seen from the example of semi-wild horses, women were not routinely receptive sexually. After insemination, no subsequent outcome hindered the male. Obviously, the menstrual pain and discomfort, followed by pregnancy, severely handicapped women, engendering hormonal and psychological changes that presented a wholly different perspective. A different sexual perspective engenders a different emotional perspective.

As discussed, child bearing and the physical demands of child rearing are enough to change behaviour and suggest mental development of a more elaborate and sensory kind; the female, of necessity, needs more wit about her than muscle and bravura. Her experience of the pain that nature routinely inflicts may have no parallel in male experience. Even in the seventeenth to nineteenth centuries (a golden era for written records, almost the first to provide a female perspective) repeated miscarriage and infant mortality shed new light on the time-span of suffering and discomfort endured by women across millennia.

Humans departed to follow distinct paths of development very early in history. No other species 'departed' from the Darwinian route of slow evolution, only humans.

Along the female path, stretching across three to four million years of human development and migration, females developed a different psychology and a psyche unique in character sufficient to match the demands of being not just human, but a woman as well.

The pelvis and the birth canal changed radically, along with the brain cavity, hips and legs; for the human was now highly mobile, developing a large brain, and expanding across landscapes to encounter wholly new challenges.

As the female skeleton adapted to walking upright, there was little room for a new human infant to be born. The foetus develops to an unusual size, one that no longer fits the primate design. At birth the infant has to rotate through ninety degrees. The tiny, pliable body, with a disproportionately large head, has to twist to pass through the birth canal with its 'new' oval shape. The new, upright gait of the human means there is less room between the pelvic bones. So the baby twists and is born facing down and backwards. In the primate world, young are born facing forwards –face up – a more beneficial position for maternal contact and putting to the breast.

So commences an extended period of weaning and upbringing that the female must cope with while continuing a complex life, arduous and tiring, stressful and far from routine. Whereas the male's mental and psychological endeavours are honed as they develop along well-defined lines, those of the female are more mediated, born of instincts honed by practical need and the presence of inevitable discomfort. These differences shape the sexual psyche of women in robust acceptance and the ability to deal effectively with their crucial place in the process of human evolution. Their labours perhaps engender even a different sense of time.

Unlike men, women's travails arose from more divergent needs and a more fluid social dynamic. So arose, we must assume, their sensory intuition. Women appear to be capable, at the same instant, of sensing, acknowledging and registering a bewildering variety of circumstances that defy categorization or evaluation.

Women possess a psyche independent from, and inaccessible to, the male. The psyche is not penetrable by 'linear logic'. Sensory intuition has evolved over millions years of human development and survived intact, despite migration across vast swathes of terrain and different climates. That it has done so speaks of its value in human society and the need to preserve, develop and respect female abilities. Women have no need to 'ape' men.

That males have always regarded women perhaps with suspicion further underlines the validity of this argument. Probably they have never 'understood' them; women don't fit into the artificially ordered world of the male brain. Their linear logic fails to penetrate the female psyche. Their reaction is to dominate and demonize. Their unspoken fear is fear of women, fear of the unknown - the inexplicable - is fear of nature itself. And as we know, Nature in the male view is a state be conquered, de-mystified, tamed. The male must conquer the female. Was it Christ who said "I have come to destroy the female" and that sex was "the act of death"? Typical male.

Rivers and streams spring from the underworld; new growth shoots from the earth; milk issues from the woman's breast, miraculously, to feed her baby. From a mysterious inner world, new life emerges.

The mind of woman is also a reflection of these mysteries, with knowledge withheld from the male. No wonder women are feared, denigrated, persecuted, and humiliated. As a sobering token of male confusion, not long ago (just a few hundred years), the Virgin Mary was venerated as Goddess and as the perfect personification of woman. At the same time, women were being burned at the stake.

Infancy, childhood, puberty, illness, adulthood, childbirth, aging, growing, dying – each these states forms a continuum for the female, but logic hates imperceptible change. It defies description and measurement.

The male consciousness thus creates its own, closed view of the world and the world of women; events only submit to observation if isolated, the mystery must be resolved, logic and action imposed. There must be, in the male view, a 'final answer'.

Within this world – the human niche – females live in a male-led hierarchy and so must compete, one against the other, for 'male' attention; the hierarchy is masculine: that is where the power lies, the power to protect. For women protection is key to their identity. So the power is the 'attention'; that of an individual male, preferably powerful. Or the attention of an imagined male, a conjured male presence, as on the cinema screen, or as presented by the media. Women must compete with each other to present a synthesized female persona that evokes a need for attention. A deadly cocktail.

No wonder a woman's self-respect is under threat. Is there any way to preserve female identity when it is no longer in their hands? So much imagery exists to create an ideal woman, one sanctioned and skewed by masculine values, that women from an early age compete against one another to fulfil an ideal. That ideal is derived from a narrow male viewpoint, at once sexist and sterile.

This viewpoint, conjured up by media industries, presents idealised males (usually aggressive and unthinking) and idealised females (thin, with child-like faces, clothes and hair, needing attention and protection), mass-produced on screen to satisfy an inherent wish among the audience: if you match the stereotype, you will be rewarded and endowed with a fictitious status and identity.

Notable is the absence in this formulaic structure of any role that celebrates fecundity, motherhood or values evoking family, children or the elder generation. This truly primitive mindset dominates human culture worldwide. Males, meanwhile, are seduced by the powerful image of a 'Kalashnikov gap-year' – travel to far-flung places and kill people for whom any human identity has been conveniently erased.

In the role of pseudo-competitor, women abrogate their true identity (by 'true' I mean primitive). We have convincing evidence now that males are indeed still the primitive creatures of 4-5 million years ago. All that has changed is the existence of a 'pseudo niche'. In 1998 this definition was first stated by Laland, Odling-Smee and Feldman in their examination of the effect of beaver dams constructed in streams and rivers; their artificial ponds created a 'pseudo niche' for other species to exploit, including birds, bears, aquatic and plant life.

By contrast, the human niche, now also a pseudo niche, arrived with the advent of recall. It consists of tools and processes but above all, language. Language plays a major role is the exponential nature of the expansion of this human niche.

"Language ...is one of the most distinctive behavioural adaptations on the planet. Languages evolved in only one species, in only one way, without precedent. .." The Symbolic Species. Deacon,T.W. 1997

The pseudo niche becomes a unique mental niche; we can access and occupy the niche – extracting, storing and using information, the key ingredients of human development. This is why humans can afford to stay 'primitive'; the mental niche expands exponentially - the more we use it, the more we encourage its growth. But the content is abstract, involving no physical development. The primitive model still lives in the physical shape of male and female humans, but the mindset in both is different, changing in accordance with the expansion of the mental niche.

AS we have seen, the human identity, both male and female, might be in danger of being lost as core social mediators erode. Mediated behaviour, once a characteristic of life in a natural environment, is now redundant. In a world full of risks and dangers, social conscience, mutual trust, reciprocal kindness and respect once played a vital part. Today the limitless scope of the mental niche dwarfs individual values. Unable to navigate such social complexity, primitive drives emerge uncensored. No physical or physical restraints exist to prevent our human deploying an infinite array of powerful resources, like the internet, the jet plane, a high-powered car, hitherto available only to privileged and uniquely powerful individuals.

Now the human does not so much exploit the niche, rather the niche exploits the human, so powerful has it become. These fundamental shifts in behaviour go largely unremarked. These structures so dominate human behaviour there's little chance of redressing the balance. Individual identity now counts for nothing and notoriety takes its place. No wonder female identity is submerged in a welter of male-orientated 'tool-using' technology; the primitive stone axe is replaced by the mobile 'phone and the internet, the DVD and the home cinema.

Language persists as a social identifier; how we speak and what we say is a signature of identity. But this is the medium of journalism, broadcast culture, politics, religion, commerce and science. Many downplay both accent and vocabulary, now signatures of a different caste; even children drag their educational feet.

As universal language engulfs the various disparate cultures of the world, male-orientated technology and culture will diminish female identity further, her dynamic intuition and her inborn sexual psyche, her voice, will vanish. A woman's character has never depended securely in combative language, instead mostly relying on translating her psyche and personality into mild protest and complaint via the novel and other 'soft' literature. She needs a voice that can rise above the hubbub. Can the female voice survive the age of the video game,

TV sport, conflict-politics, large-scale centralised commerce, instant world-wide messaging? Subtlety and nuance, inference and tone of voice, gesture and facial expression are important female communicators, not to be disparaged, diminished or dismissed. Women need a special means of vocalising, in every sense, before they are condemned to silence by the sheer weight of technology.

In England, in the last century, rather than demand the vote, suffragettes should have demanded a separate parliament.

Women could in this way have cemented their identity. They cannot do so in a male-mediated social structure. This radical change would have been better than the slow absorption of the female sexual psyches into a technological patriarchy. I hope to show that it is not females alone who are at risk of being permanently captive to an increasingly narrow and more confined social path.

CHAPTER 8

The Human Infant

In 2005, and to the joy of researchers, closely resembling breathlessly expectant parents, an infant proto-human, a 3,300,000 year-old baby girl, was added to the scarce caché of remains that comprise our human history. She was immediately christened by her discoverers 'Lucy's Baby'.

At last! Lucy is one of the earliest hominines; a broad species group that also includes modern humans. Lucy belongs to the *Australopithecus afarensis* group, one of the earliest ancestors of modern humans. Her few bone fragments date to around 3.2 million years old. So human types have been around for a long time; and, of course, so have their babies.

Babies are important in anthropology and palaeoanthropology; in the case of the latter they are very, very rare. And the skeleton of 'Lucy's Baby' was pretty much complete. She has overshadowed 'Lucy', one of the earliest examples defining human lineage, igniting debate over how and when human ancestors adapted to 'life on the ground' after 'life in the trees'. You can join the fun: find a baby and stick out your finger somewhere near its hand. Chances are it will grasp it with a surprisingly strong, assured and confident grip, not even looking at you, or what it is holding onto – it is instinctive.

In an incredibly short time, each baby goes through an evolutionary journey that mirrors human history, right back to Lucy and her offspring and beyond. It begins – well, we all know how it begins (though early humans probably didn't), - but once fertilisation takes place, development of the embryo and the tiny bunch of cells that represents the growing foetus is a mirror image of our evolutionary history. Of course, this is hard to examine and difficult to replicate. So, in the case of the 2005 find, the actual skull, limbs, facial bones, ribs, shoulder blades, some parts of the legs knees and feet, represent a treasure trove. Like all babies, Lucy's baby has stolen the show, as babies always do.

The modern era re-enacts a family story that is, quite literally, ages old, even 'timeless'. Today it might go like this;

"Hello?"

'Guess what Mum?'

"Oh, it's you Mike. Nothing wrong is there? How's my Jilly?"

'It's all over – four o'clock this morning'

"What! Already? So soon. Is she all right?"

'Jilly's fine. The baby's fine'

"A baby! Oh, Mike! A boy or a girl?"

'It's a *baby* Mum!'

Notice that Mike didn't stick to the script; to him the sex of the new baby is secondary, but this would be very rare. We are typecast from the word go, or even beforehand, given current scanning procedures, to be intimately concerned with a baby's sex, desperate to know. Mike is probably teasing; pretending to be a bit right-on, new-liberal, non-sexist, equality-obsessed, new-age dad, but absolutely tickled pink. He most likely knows already ; it's a boy.

From now on – from the hormone-laced environment within - another life begins without. There is Mike to deal with, and Jilly, their hormones and sexual psyche. This experience is one for which the infant is already primed. It has to soak up as much possible, as fast as possible. Here we need not concern ourselves with the growing child. Every possible authority, from neurosurgeons to everyone's granny, has dissected this subject at length. However, there are unseen, unrecorded aspects to early human life that relate to the peculiar position of the human in nature's scheme of things.

What has gone unnoticed is that games with infants carry an intellectual signature that is uniquely human; the game of 'peek-a-boo' is but one example, a game both child and parent love. Take and object, place it in the child's eye line then make it disappear from view. The child looks blank until the object reappears; 'peek-a-boo!

The sound is a 'cue' and a stimulating one. Children learn fast – they have to, there's a hell of a lot to learn. The point about the game, though, is not that the object disappears and delight abounds on its re-appearance. The point is that the child becomes conscious the object is somewhere else; even though it is not 'in sight', it hasn't ceased to exist. 'Out of sight, out of mind' is not something that applies to humans, even very young ones.

The object is still 'somewhere', and this is a crucial facet of human thought, instinctively cultivated by parents. That's why they enjoy the game, too. It is vital that they have the instinct to pass on that sense of enjoyment, vital that it carries the ingredients enabling the (still very young) infant to begin making abstract connections. Vital, because it's the signature of the learning process itself and it has to start right away. They are handing on a uniquely human skill, holding 'in mind' an event in the past to 'create' anticipation. This is recall, which, as we have seen, is the key differentiator between human and other animals.

Children become loved, adored, prized, spoiled, the centre of attention. They embody all the potential skills of the human. We delight in seeing consciousness emerge and *our* expression of delight is also a key stimulus for the child. It encourages better, more varied and more finely-honed skills. To see just how much we value our brilliant offspring, reflecting of course the characters of their parents, grandparents, uncles and aunts, here is a slight diversion; we're off to Monaco.

Because real estate is so expensive, Monaco investors and planners go to subterranean lengths to satisfy the demand for new commercial space. In one underground shopping centre, complete with chandeliers and marble floors, restaurants and smart cafés, the firm of Gucci has a shop devoted to infant fashion. Here the indulgent parent can put its infant in beautifully styled shoes before it can even walk. Of course, in a month or so, that pair is redundant. But parents, if they can, will indulge their offspring with the best that money can buy, perhaps little realising that the best thing they can do for any baby is repeat the simple games of infancy they learned at the very beginning of conscious existence. Perhaps 'peek-a-boo' is more vital than 'Gucci-Gucci-Goo'.

No. Shoes in blue for boys, spangles optional, shoes in pink for girls, patent leather or suede, is the colour-coded tradition to which parents rigidly adhere. All parents and grandparents 'spoil' their offspring; that is they attend to them, stimulate them, encourage them, endeavour to guarantee their love with gifts or attention that sometimes seems excessive. This is a modern habit; a massive choice of products simply exploits this natural response. Not long ago there were no products for infants and few for children generally; they made do with hoops, string, conkers, imagination, rhymes, stories, playground friends and 'pretend games'.

The 'Monaco effect' is now apparent in many town shopping centres and supermarkets. Even market stalls carry spangle-bedecked high heels and stiletto shoes for late-infancy children. The obsession with juvenilia permeates both young and old and the process of maturation seems to be increasingly delayed or even failing. Now, young people and even adults adopt a babyish manner of talking and behaving, following years of encouragement from parents and elders and an even broader cultural acceptance that adulthood is no longer a desired state, childhood is here to stay.

Anthropologists emphasise the longer period required for human infants to mature when compared with animal or primate offspring. Palaeoanthropologists single out the size of the skull cavity of pre-hominids to denote advanced or limited mental ability. In this they are wrong, of course; an early precursor of humans, an orang-utan, a chimpanzee or a gorilla has exactly the right number of millilitres in terms of skull volume to be a perfect specimen of *Australopithecus afarensis*, orang-utan, chimpanzee or gorilla. However, in order to mature into a well-developed human in the 21st century a long time is needed and the amount of time increases with every passing year. No wonder baby-like sounds become acceptable in adolescent and adult speech.

'Teenagers ; they are still maturing', is a common-place truism.

Our new family treads a risky path between power and excitement, sympathy and restraint, independence and 'selfie-obsession', concern and empathy.

Much of the character of a child, male or female, is laid down prior to birth via   epigenetic transfer. This is the influence of external, arbitrary factors on the infant across or within the placenta. Obvious examples are the mother's diet – in *Epigenetic Inheritance and Evolution*, E. Jablonka discovered that dietary habits of the mother could influence the later preferences of the child for food flavours. (Journal of Evolutionary Biology,1998)

More work done since, discovering the importance of cortisol, stems from epigenetic research. This suggests that mothers under stress may put at risk their baby's developmental health or wellbeing. Epigenetic studies also confirm that the amniotic fluid is an active, rather than a passive part of the evolutionary process; developmental changes are passed on to subsequent offspring via subtle genetic changes, beneficial or detrimental.

Thus, mutation and any resulting subtle genetic changes are not solely the result of external forces (i.e., from the physical environment). Social conditions may play a significant role. This may include, diet, medication, alcohol, drugs, noise (causing stress to the mother), air quality, emotional balance or imbalance.

None of these subtle influences compares with the mighty power of a genetic structure that first evolved over 4 million years ago. More important than that, human genetic makeup *includes* the genetic imprint of ancient species from which we, in turn, have evolved. We are the descendants of the first organisms that appeared on earth; we go back to the origins of life itself, right back to the primordial forms seen in the developmental stages through which the human embryo must pass.

DNA structures are thus robust and extremely long-lived. While we are aware of our relationship to chimpanzees and other primates, we should also remember our shrew-like and fishy ancestors. There may be something of a family resemblance somewhere.

So our new daughter has an impressive pedigree. More impressive is the sex-typing discussed before. That we took to male/female division of labour so early helped our evolutionary success. It probably also spelt the demise of close hominid relatives; we simply outperformed them, out competed them for food, water, shelter and a share of the available food.

The evidence is there to see in the distinctly different behaviour of infant girls and boys. Sensory intuition is there in a baby daughter. Linear logic is there in the boy. Show a daughter a doll and she begins to relate to it, even early on, recognising and absorbing the characteristics of face, appearance, attitude and dress.

Boys love action toys, sticks, balls and moveable objects, indeed objects the whole point of which is 'action at a distance'. Naturally, parents encourage and reward what they see as appropriate behaviour. No doubt this is pre-programmed, too. Women also want a female companion. Men; a male companion, even if they still have to 'grow a bit'. Perhaps an element of re-enacting our experience of childhood means we pass on 'appropriate' behaviour. Perhaps this, too, is partly the reason for our evolutionary success.

Children are fascinating, not just to parents. Perhaps mentally they also embody a mirror of the time sequence that was early mental evolution. Is there a subtle point in infant development at which we can recognise the evolutionary equivalent of the recall event? There *is* a period of vacant staring, during which animation is present but where attention is almost absent. This state seems to be rather like that of an animal; everything works but the world has little significance. It makes sense to us but not to an infant. We are impatient to see responses that are recognisably human.

This seems to occur when the infant is first able to follow movement in a consistent manner. This is the 'gaze'. As soon as this ability is established, consciousness probably arrives, though this may be impossible to test; we don't know where consciousness or recall resides in the mental architecture, much less how to test for it in a baby. Boy or girl, the differences in behaviour begin to emerge. The very primitive division into two psyches or personalities becomes apparent. Preferences for toys, food, people and games begin to make themselves felt.

As soon as children become mobile, development accelerates, with boys more adventurous, often seeming to be less aware or concerned with risks. Statistically boys are accident prone to a greater degree than girls are. The NHS Health Development report of 2005 reported that Boys suffer around 30% more injuries than girls in the 0-5years age group do. However, a curious statistic from another source (RoSPA) shows that far more young girls than boys consistently suffer incapacitation through 'acute over-exertion'(perhaps trying to catch up with the boys).

Boy	Girl	age 0-4 years.
1958	1454	Same-level stumbles.
1107	907	Stairs
24	14	Ladder or steps
30	36	Fall from building
3209	2440	Other fall
688	428	Collision with object
111	77	Puncture wound
206	342	Acute over-exertion

Many of the injuries (not listed here) are burns and scalds, but these are home injuries. Often children are injured are killed in road traffic accidents. The car is now more dangerous than the home, especially as the children grow, and the catalogue of other injuries lengthens (especially to boys) as they become more adventurous and confident. Suffocation, poisoning and drowning occur, also in the home, in the garden and as children widen the scope of their exploration of the outside world. It has always been a dangerous place for the young in contemporary times as well as prehistory. Children have survived all kinds of disasters.

Mothers suffer too, despite their vigilance, with an increasing number and variety of dangers. These now take the place of known and unknown diseases of the past; septicaemia, viral and bacterial onslaught, pneumonia, plague and 'sweating sickness'.

Nevertheless, children survive as they always have; the population increases despite lower birth rates. This may not be the case in the future as antibiotics lose their effective edge through overuse. Humans developed antibiotics to combat bacteria. Paradoxically the longest surviving organisms ever, bacteria themselves were present at the very beginning of 'life', over four *billion* years ago, long before the appearance of even the most primitive of our ancestors. We still have a long way to go.

As puberty sets in, massive hormonal changes account for what seems to be churlish behaviour, withdrawal, shyness and confusion. Then there are spots.

133

Considering the emotional and chemical upheaval, this is the worst time children should go to school, unless this serves as a good way of introducing them to disease, thus 'firing-up' the immune system, in the meantime bringing home the bugs they've caught. While at school, concentrating on arcane subjects takes a lowly place on the agenda as boys test their mettle within new male groupings. Some girls suffer the agony of unfamiliar emotions and hormonal highs and lows. Sexuality shoulders aside society's carefully planned programme of education, while mental development struggles against the physical onslaught of 'growing up'.

Unfortunately, the early division of labour and the particular skills, taught by ancestral custom, at which sons and daughters would have soon become adept, are submerged by organised education. Early customs of boys becoming boys and girls becoming girls served to maintain and emphasize the division of the sexes, until perhaps a more confident time, when sex-differentiated skills had already been acquired and the onset of maturity apparent.

In the modern context of mixed schools, the constant, distracting presence of the opposite sex, at a time when intellectual effort is supposed to displace the physical, is not a happy time for young people.

Parents are no longer confident, either, in laying down guidelines. Responsibility assumed by the State promotes a disconnect between parents and children, their social mores and local society. The patriarchal role of the State inevitably diminishes the female role. With its historically male values, its elaborate linear logic in the learning process, its standards and rules, the educating machine fails to take into account the more subtle nature of the female; her sensory intuition, her innate skill in establishing natural bonds within the family and society, the ability to preserve, with other mothers, resilient and enduring local communities.

It is true females do well at school. University intake is higher than males. Boys suffer at school and fail to engage. They are more easily distracted because the male nature is far better at solving practical problems and physical challenges . Schooling methods demonstrate again that separation of the sexes makes for more effective education and better life chances for both. Boys need the discipline and influence of elder males peers they can respect. Females need calm, secure environments away from pushy, aggressive boys. School is no place to add a toxic hormonal mix into the environment of the young. The large- scale educational establishments with a preponderance of beleaguered female staff who fail to connect with young males perpetuates ignorance, aggression and anti-social sexist behaviour.

Nothing in a patriarchal State considers how to encourage family cohesion or continuity. In an age when grandparents and the elderly are lonely and feel useless, their children and grand-children become separate and autonomous social groups by perpetuating childhood. Statistics and the logic of State divide cohesive families into separate social entities, by education, by age, status and by cost or value to the exchequer; the currents of organisation divide families and diminish female influence. Paradoxically, mixed schools divide the sexes further.

The young cannot appreciate or respect the nature and values inherent in the emerging sexual psyches of their counterparts. Forced too early into each other's company by a patriarchal State they have no chance to mature or grow into their own identities.

The State does not intentionally undermine female influence in society. But it does rely on the male sense that chaos is the enemy of organisation. Education is formulated along these lines: timetables, regimentation and control. Paradoxically, the female is associated closely with nature and thereby with chaos and the chaotic nature of the world. Men instinctively sense that chaos lies at the centre of nature and is the enemy of all their purpose. They fear femininity, as far as they can grasp it. Women are a seen threat to the male even at an early age. Even juvenile males will make life hell for young girls. Chaos must not reign.

Males confronting 'chaotic' nature is what young males instinctively wish to engage in. Learning 'male' skills is what they yearn for. Education fails them completely. No wonder their scary school companions do so much better.

Meanwhile our new male will grow fast, if he survives the risky business of babyhood, in the benign environment purposely shaped for his success. Physically, he is as primitive as his million-year-old predecessor is, in that warfare and conquest attract him most. New video games satisfy the visceral thrill of death and injury inflicted upon a myriad of challengers to his masculinity and territory. Society has found a way of training its new recruits to the male hierarchy while minimising the danger of actual physical injury, even if these puny new humans are no match for the real thing. To the young male, war games are electrifying.

The concomitant solitariness, the lack of adult guidance, lack of adherence to the rules of society, the thrill of primitive behaviour; the State has a compliant, malleable citizen ready-shaped to conform, once the right buttons are pushed. Does male education produce toy soldiers, ready for the virtual world of work, with its mechanised decisions and responses? Is this the intended legacy from the State; that the young male inherits a taste for power and material gain?

Certainly, the birth of a son to a woman is unlike the birth of a daughter. The former, a little kernel of a different kind of power or ambition, the latter, an ally, a co-conspirator, in the matter of sentiment and personality, of companionship. Some women must wish for a son, just as some might prefer a daughter, but at some point encounter disappointment. As soon as she knows the sex of her off-spring does her behaviour subtly change. What effect does this have on the new-born.

That a girl might be favoured *after* a boy has been born (a sort of back-up copy) reinforces the sense that a family is secure. It may be that another ancient precept makes its presence felt; a male reinforces hierarchy, promises added protection to the mother, a kind of second fiddle to the father. Perhaps, unconsciously, both male and female sense that a son implies multiple offspring elsewhere, with other women, extending the influence of father, mother and family.

Carrying different promises or outcomes at the time of birth is implicit in the sex of the child, so sexual psyche becomes established even as, perhaps before, personality makes itself known to the parents.

The birth of a female child may comfort a mother but make a father anxious; he may want to know a male will be forthcoming. The permutations are endless but they all involve the introduction of additional sexual psyches to the dynamic of a developing family.

What is certain is that within a short time of being born, like her brother, the sex of the newborn resonates. Pink-coloured little fashion items are already in evidence if granny has anything to do with it. The questions of names is of vital interest and (as we've already seen) can have lasting consequences. Again, circumstances during pregnancy can influence later behaviour in a child;

confronted by stewed apple or mashed broccoli she may have no problem.

She may have the notoriously famous 'sweet tooth'. A later appetite or complete distaste for alcohol can appear. The science is poor, so folk tales reign supreme.

Under normal circumstances, we have to assume that, like her brother, sexual characteristics become more significant as she grows. While her brother did not take readily to frills and sparkle, his sister will. She'll demonstrate (to brother and father) curious and baffling shifts of temperament and rationale, making connections that seem to them bizarre and disconnected. Emotions already carry more weight than male 'reason' in forming relationships, even if this involves a garden worm or a discarded sock.

From a growing female perspective, father and son, even at a young age, display characteristics that seem too physically orientated to be healthy; noise, speed, brutally direct attempts to extract responses from their surroundings, all these interactions with the world appear odd to a more contemplative, more intuitive female. Aggression fuelled by muscle power from her brother is something best avoided. Strategies develop to divert attention as soon as she can master them. Expressing hurt is one she might find encourages protection. Meanwhile, her father encourages her brother to make light of pain and discomfort.

From these strategies evolve specialisations at which women are best; instead of combating the world 'head on', attention to circumstance, cause and effect at a detailed level avoids muscular clashes with reality (we've already seen the accident figures). Thus, the fine detail of her surroundings takes on a new meaning. She realises this 'other world' is neglected by males. She can make it her territory. In exploring it she comes to feel its significance.

Life takes on a more secretive and personal slant. Her mother shares these sensory and intuitive perspectives.

Our little girl is growing out of her shoes, and, with every new size, an increasing myriad of styles. Her virtual world – away from the turbo-charged hormonal whirl of school, on-set of periods, sexual exploration and experiment - is a personal haven; a concentration of femininity.   She is encouraged by education (male orientated) to subdue sensory intuition in favour of a more (male orientated) consumerist and material world. In an electronic environment, sensory intuition is neutralised by a far greater pressure; to become a consumer.

A sensory world - environments in which women could explore and develop ancient skills – hardly exist. Nor would it receive general approbation. In a mechanised, industrialised, electronic world, sensory intuition perhaps has no place. Only the commercial world exploits her sexual psyche, recognised not for its origins, or its real nature. Retail shops, and especially the supermarket, mercilessly exploit the ancient gathering skills. If only women could or would push larger trolleys, profits would be even greater.

Inevitably, the fate of young women is, after glamour has lost its glitter, as a consumer. The ancient detail of the natural environment recreated on the shelves draws her in. The choices are endless, with no means of sensibly navigating them. The dilemma for the maturing woman is the realisation that commerce, engineered by male linear logic, is actually at odds with her nature and her skills. She can only 'buy' or 'trade'. The social, political and economic dynamic is male. Sensory intuition has no place. She must adopt a male persona.

Female instinct may be as visceral, profound and constant as male aggression. But sensory intuition is contrary to modern precepts: competition, constant renewal by the demise of the old; challenge, continuous renewal, fresh choices and new conquest is the continuing impetus that fuels patriarchy.

By sleight of hand (again unconscious), patriarchy trumps matriarchy; consumer mechanisms are perfectly shaped to provide the palliatives; even for five year- olds there are 'glamorous' shoes, handbags, dresses, lipsticks, hairstyles, all the trappings of an early 'adult life' from which the 'natural adult' quality has been so skilfully removed, translated into 'goods', personal taste neutralised, suborned by mass production. The artificial experience is now valued more highly than the 'natural' or the 'original'. Five-year-olds mimic twenty-five-year-

141

olds and thus immediately are ready for market exploitation. The male hierarchy wins again.

Thus youth is prolonged, well into what used to be called 'marriageable age' and for as long as possible; the individual becomes so enamoured by the idea of the State as parent that it is becomes easy, indeed essential, that individual responsibility, freedom and ideas are abrogated. Commerce is awash with material rewards and endless choice. No individual can compete. The sensory intuition of the female is overwhelmed. The male's energy and inventiveness is made redundant. The mental niche rules the human, not the reverse.

The State looks on while the rout of individuality, and especially female personality, continues apace, much of it through ignorance and missed opportunities. The Sate condones diluting individual worth because economies depend upon it.

Our growing girl would benefit most from a strong, coherent family. State education, with early enforced mixing of the sexes, adds a level of distraction hard to avoid. Pressure to 'do well' and choose a career path is a constant strain even at an early age: parents already know the exams she has to aim for. They share her anxiety. A young person's idea of why the pressure is on is hazy, at best. Such strains arrive in the family, for males and females, just when they need time to grow and when a calm atmosphere is paramount.

Schools and universities are not the carefully structured and sympathetic institutions they need to be. Teenagers and young adults will always find a way to 'mix' with each other. What they may need most is less distraction in a world that is daily more complex and pressurized.

Many adults, looking back on their life at school and college, have a great sense of regret – they wish it had been different. Most males would have wished they had known more about females, and *vice versa,* and much earlier. So, perhaps education should have started with sex. Most would have perhaps wanted 'life' and education to be separate experiences. If it had been thus, they might have been better at both, enjoyed them more and avoided many more pitfalls.

The male has his other competitive world, one that wholeheartedly, resolutely supports his psyche; he and his brothers-in-arms have built it. This may be the reason why sex education is not on the curriculum.

Ever-more complex societies rely upon populations that are compliant and co-operative.

One might think women have suffered enough without having to accept that final indignity; to become surrogate males.

The ancient mutation that gave recall drives a male patriarchy because it works so well; the mental niche expands exponentially, offering endless novel ways to exploit the resources available. The male is rewarded .

The suppression of a contrary view; one of husbanding resources and of renewal, runs counter to an accelerating human context, a pseudo niche out of control, for there are no natural limiting factors except ever-increasing consumption and finally, exhaustion. Here there *is* a female role, and we'll come to it later.

First, let's examine the subtle dilemmas that women have to cope with everyday, the way they learn to exploit them, while successfully retaining both allure and identity. But there are risks involved.

CHAPTER 9

Patriarchal Society and the Abuse of Women

Let's look first at women's legs, notably that indefinable margin between good taste and outright vulgarity, the hemline. If a woman is skinny and stays upright, the sky's the limit for the hemline,. Every photo call and fashion shoot exploits the opportunity. Seated, all sorts of technical problems arise. Female newscaster's deal with the attendant problems demurely. However they rarely wear trousers or a longer skirt. The broadcaster exploits any minor frisson of excitement mercilessly. TV is a medium steeped in commerce first and foremost. The product is one that efficiently exploits, encourages and perpetuates such responses. Do women mind?

The shadow cast by a skirt over the knees exactly mirrors the pubic area in its more natural form. Women might be more cautious if they were aware of the origin of the frisson so broadcast. I hope not.

Are women aware that the constant use of lipstick draws attention to the fact that labia are powerfully suggestive, especially in a range of such delicate shades of red and pink? So many devices and mannerisms exist to perpetuate and underline the persistence of femininity and its signalling devices that we can read them in parallel with other species. They are employed as a natural ingredient in ensuring continued survival. Long may it last. It is curious, though, that the female in naturally discreet. That men need so much encouragement is due to two factors. One; they are basically dumb. Two; the power of recall. Men are so preoccupied in a world of linear logic that they need a surprising amount of help.

Rape statistics – to the surprise of no one – are notoriously unreliable. They are absent, or distorted, unavailable, censored, altered, unrecorded, lost. This is the universal story across the world. There are societies where rape is not a crime or even an injustice. In some countries, it is a truism that women provoke men to rape, the implication being that the man is not the guilty party, rather the victim.

In 2015 the incidence of recorded rape shot up by 31% in a year to 24,043 cases, the highest for 10 years.* Of course, recent successful prosecutions encouraged victims to report crimes, rather than stay silent. This improvement may be welcome, but ignores the year on year suffering, the pain from untold past years.*(ONS. Reported in the Guardian. 8/12/2015)

The figures do not point to an increase in rape. Rape victims per thousand of population decline in number from the 1970's: 2.5 – 2.8 – 2.5 - 2.3 – to 1.3 (1994) – then 0.9 (1996) – 0.6 (2000) and in 2003 0.5 per thousand of population.

Is this is another statistical piece of sleight of hand? Should that 'per thousand' proportional figure be doubled, assuming half the population (male) was not a victim of rape? Using proportional figures for the entire population mathematically diminishes both apparent incidence, frequency, and thereby the seriousness of the offence. Figures for rape of women are better expressed, not as a proportion of the population, but *per thousand women*. The figure is thus 1 in 1000. Either that, or a woman, revealingly, is actually a 'half' person for statistical purposes. Figures for recorded rape are imprecise or misleading. The real figures are probably much higher, because most offences go unreported; the rapist is often a husband or close family member.

The incidence of rape in society appears to mirror the degree to which a patriarchal religion dictates custom and law. In extreme cases, women are mere appendages to male status. Some males believe they have the right to abuse women as they wish. Even unmarried men escape assault charges even though the women of their culture are perfectly modest in attire, demeanour and behaviour. To men – behaving primitively, on instinct, through uncontrollable urge, sex is gratification. We can't say 'pure' but it is certainly 'simple'.

For the male, few consequences ensue; legal, moral, familial. For the woman the consequences are profound; physical, mental, emotional, moral, medical, long-lasting, permanent. A man easily forgets a sexual encounter. A woman cannot, especially if disagreeable and forced. Her sexuality is her persona. To violate it violates her entire person.

I suggest this anomaly relates to the description in a preceding chapter; sexually aggressive men ('if they can get away with it') are yet fearful and uncomprehending of women. Women undoubtedly have a power over men which, in less amorous mood, they reflect on and perhaps resent. Because sex for men is a more transitory and shallow experience, earth-shattering though it may be, there are no lasting consequences. More crucial and mysterious is its power to overwhelm their assumed physical strength, status and bravado, throw into doubt their entire sexual psyche.

For however brief a time – sometimes very brief –sex reduces them to jelly. This is the nature of ejaculation, the vital firework thrown by nature. Men are shocked by the realisation of their transient significance. In a sense, the sexual contribution is made almost as a reflect action to the allure of sex. Yet a simple but powerful response to a the presence of a female 'lights the fuse'. For men , after all the fireworks, there's a sobering realisation that they are of minor significance in what follows.

There is much more to woman than meets the eye. An internal power is in her that eludes men. She seems to be 'nature incarnate', having a deep and meaningful relationship, not just with her mate, but with nature and the ensuing process of pregnancy and birth.

In his unvarnished role as dominant male, Nature appears as something odd, mysterious, unfathomable. Just as in physics, engineering, astrophysics, whatever role he chooses to express domination, his entire evolutionary history is aligned to conquer the forces of the natural environment. Nature is in this sense seen as resisting him, to be shaped, to be mastered and controlled, bent to his will and design. Thus, male attitude is enshrined in patriarchy; losing dominance robs him of the only thing he has – *his* sexual psyche.

The more that women strongly represent the power of nature, the more that men are discomfited. Naturally, their linear logic is under threat. Worse, the *hidden* power of Nature exacerbates the case; the more subtle and mysterious, the more elaborate his sexual psyche, the greater the threat. The greater the male need to defuse female power in the light of his uncertainty, the greater the threat of rape and abuse of women.

A sad paradox is that those religions supposedly most concerned with purity of soul, with behaviour and ideas spirituality inspired, with a profound knowledge and respect for long-held beliefs, with reverence for long-accepted custom and scripture are led by men who fear the power of nature most. Their piety is a measure of how far patriarchy is divorced from reality. Under conditions of strict religious observance are those most likely to abuse and humiliate their women those with the strongest conviction?

Abuse and humiliation is in reality a denial of sexual identity. Women and girls suffer confinement bordering on imprisonment, lack of education, have no chance to voice an opinion or even the freedom to show their faces.

149

These religions, professing a search for enlightenment, condone by silence and enshrine by ignorance a blindness to the beauty, intelligence, humanity and insight of half the world; its women.

It was true of Catholicism and Protestantism in early European history and true of many religions in which a core belief is that 'God' is male. A demonstrably false conclusion in that 'man' is only the most recent, and probably the most seriously flawed example, of a process of speciation dating four billion years before the arrival of the 'human'.

It's probable that, throughout the history of humankind, since the adoption of manifest tools by the male for use in hunting and warfare, women have come off worst. Before that time we can only speculate. And speculate we will.

While it's perfectly reasonable to assume women certainly had a hand in making the first tools and probably took the lead, men gained the upper hand through hunting experience and their ability to provide large amounts of rich protein. The co-operative effort necessary proved the power of the male group. The rest is history.

Stories of female-ordered societies of any scale or duration, while engaging and poetic in the implication that such societies were thereby more pure and closer to nature, are hard to verify. Perhaps such stories are wishful thinking. Even transient female heroines are celebrated and admired, speaking to a deep-seated need for a more neutral balance. Men have always been trouble. Boudicca was trouble, but female heroines have a special place.

The probable truth is that recall in humans – an evolutionary 'flaw' from a Darwinian standpoint – while bestowing the ability to modify the environment at will, contains no counter balance, no modifying influence; we (men especially) go on improving Nature for our own specific benefit. Females hold a place closer to nature and perhaps we seek in them a means of redress. Females 'go along' with male dominance because, as in 'Nature' (as with other sexually ordered species) there are positive benefits. There are negative influences, too, as we'll see in the following, somewhat harrowing, chapter.

In this regard, humans display the classic profile of any animal species; a dominant male is better placed to mate with the best possible example of the female. In turn, the female benefits from those genes that will confer on male offspring the genes of a dominant male (one that gets the best pick of the nourishment and mates). Female genes are passed on to female offspring; they are choosy, perhaps the most choosy, and that's all that's needed.

A choosy female is nature's choice. It's in her genes because her mother passed on her genes. These 'choosy' genes led her to choose a mate with character enough to pass on the best chance of survival thus repeating and refining the character (the genetics) of previous generations. The best genetics are determined by the niche – that bit of nature that does the cold calculation. It is the niche that 'sorts out the men from the boys'. The individuals least well-adapted to exploit the special parameters of the niche die off. The special parameters of the niche (advantages and disadvantages) determine, in turn, the special characteristics of the species. One of the characteristics of a species dominated by recall is consciousness, prompting us to unravel the 'before' and 'after' of the time illusion.

Our manipulation of the environment, the world as we see it and, hence, the universe, demands we supply an explanation for forces we encounter. We find the most likely explanation in a male divinity. Here we find the reason why females have been relegated to merely a supporting role; the dominance of the male, his aggression and inquisitiveness, the bravery selected for by a history of hunting and warfare.

It is impossible for men to imagine that a woman could imbue the human with consciousness, forethought and the ability to recall events, abilities denied all other creatures. Only male divinity could be their source. The dominance of the male, his aggression and inquisitiveness, inventiveness and insight make the male the prime candidate for 'Overseer' of nature and mankind. Man becomes God.

The 'purpose' or 'design' in nature come from this bit of wishful thinking. Evolution is a cold-hearted lottery determined by the physical state of the world, its moisture, its air, its seas and rocks, its forest and swamps. Behind this landscape, there is no 'will' or 'consciousness'.

It is time for humans to recognise where we came from and what we are. Having now the insight to ascribe to both male and female their respective sexual psyches, we can recognise that women have to be legally and morally equal. At the same time, we can recognize that patriarchal dominance is not only out of place, but out of time. Religions of this kind are defunct.

Women must rightly claim their place as the main embodiment of everything natural, meaningful and real. They were shaped, not by an invented male divinity, but by something more profound; billion upon billion of years of evolution.

Though it is clear that there are no longer matriarchal societies, and no armies are under female control, patriarchies control governments and societies in a manner that is, by any human standard, unjust, unwarranted, unmerited.

In 2015 Saudi Arabia allowed women to vote for the first time, though they are not allowed to drive a car (to get to a polling station, for example). They were also allowed to stand as candidates providing this was sanctioned by the obligatory male guardian (females in Saudi Arabia must have a male guardian). Women could now vote for candidates (978 were aspiring women, against 6,000 male candidates) to represent them on local councils, though their influence is limited.

Women voters comprised only 130,000 of those registered to vote. Thus, from a total of registered voters of 1.5 million (even then representing only 4.75% of the total population of 31,540,000), women represent just 0.4% of the 'voice' of Saudi Arabia. A long way from modern women's notions of democracy, but a start, of sorts.

This state of affairs in a Sunni Monarchy, is not seen as restrictive. It is 'protective'; Saudi men are seen as absolute guardians of their women and Islamic Clerics and the Monarch agree to share responsibility for the religious, moral, legal and economic stability of the country, absolute guardians of its welfare and well being. Women today are not just (with their guardian's permission) teachers or nurses, today they can run their own (modest) businesses.

 Female candidates must lobby through a male guardian to express their views,   or address male voters in a booth with a suitable screen separating voter and candidate. Custom is maintained. Robed in black, the candidate is completely covered from head to foot, with only the eyes visible.

Against this, Saudi Arabia is the world's 18th largest economy by GDP (IMF, World Bank, UN, CIA World Fact Book). The UK is the fifth largest. Despite its economic importance as one of the world's largest exporters of oil and being the world's second largest arms importer (Stockholm International Peace Research Institute), Saudi Arabia has its critics:

Ali al-Ahmed writes:

"Religious textbooks   (are) comprised of medieval ideological indoctrination"

..and .... " Denying the anthropomorphic characterisation of God by the Wahhabi and Salafi groups makes you a polytheist" (and thus you may legitimately be killed).

"The primary goal of Saudi education is to maintain the rule of absolute monarchy by casting it as the ordained protector of the faith, and that Islam is at war with other faiths and cultures"   (Ali al-Ahmed, The Guardian 2010.)

Patriarchy in this context is complete; a religious and monarchical co-operation in which only designated and hereditary male leaders steer the ship of state and all aspects of culture, sanctioned by males. The Sunni state religion is also the guardian of the image of God as male and omnipotent. But we mustn't get carried away with indignation and a distorted sense of the fairness of 'Western' values; remember the women burned at the stake in 1643. Religion in England was also extreme, patriarchal, monarchical, tyrannical. It's a question of timing within a historical context. What's a few hundred years out of a few million ?

From an evolutionary viewpoint, but unlike her dull, inconspicuous female equivalent in the natural world, the male now directly influences her appearance and behaviour, providing the evidence for un-natural selection. Much of her appearance and behaviour is dependent upon male preference within the species. So nature itself is a slave driver; 'unattractive' women don't attract 'attractive' mates.

Just as much a captive as in her natural niche as any animal, slowly consciousness is dawning of the freedoms denied to her forebears. Humanity it seems – despite our myopic view of ourselves - takes an aeon of time to become a fully paid-up member of the pseudo species, one in control of a chosen, not arbitrary destiny.

Populations are still enslaved, more by commerce than by religion. Indeed, so called '21st-Century', 'modern', 'advanced' or 'civilised' nations, also enslave their populations to a veritable tidal wave of 'pseudo bounty'. Nature's bounty, all that we needed to ensure survival, becomes rapidly replaced with stuff we consume in another, entirely artificial way. Consumption of pseudo bounty is highest in regions with high GDP.

In the U.S. the latest film releases, such as 'Star Wars', merchandising support supplies toys based directly on their warlike plots. Boys and men are the natural target. The primitive appetites for death and destruction are represented (it is assumed) in a safe, coded form; plastic weaponry, threatening masks, make-believe armour, robots – all express the still vibrant appetite for 'the hunt' in which male ancestors became so expert. Pseudo bounty is the reward.

The 'prey' is now screen-based evil space enemies, but the visceral energy is still there, exploited to the full: one male collector has amassed over 9,000 items over 15 years as the 'Star Wars' franchise (now owned by Disney Corp), pumps out episodes to a market as captive and enslaved as any in Saudi Arabia.

The 'Barbie Doll' toy aimed at females is losing its power. The female fan is still not keen on the aggression reflected in the armed and deadly space-ship plastic-reproduction business. Even the original Princess Leila reproductions gathered dust on supermarket shelves as product aimed at young females remained unsold. Disney are now learning something about their female audience. The research is coming through; motivation is still markedly female. They are having success with shirts and bags.

Just as male primitive urges survive, so do those of the female. Intuitive responses shape society less and individuals more. A sea-change in evolution is taking place as the speed of the expansion of the mental niche becomes exponential. Like Saudi Arabia (an extreme example), state control of countries which congratulate themselves on their fairness and liberality has unlooked for consequences. The influence of the female is felt less in these societies as the State takes control.

In the United Kingdom not long ago (2008) figures from Child Care Services show 802 new-born babies had to be taken 'into care' for their own safety. Within five years this had risen to more than 2,018 new-borns. 50% of these mothers already had suffered the pain and loss of a previous baby having been taken. 66% of these mothers were 'in their teens.

"There was a more general trend towards more timely action". (Dr Karen Broadhurst).

Here the state was performing a duty to care for newborn children as a 'statuary duty', as laid down in guidelines for the Family Court. But, the outcome was a 'disproportionately increasing' number of new-born infants being taken from their mothers. The instincts of the mother had been 'drowned out' by social circumstances outside her control and any responsibility – or even involvement - effectively removed.

The bare statistics hardly begin to reveal the true extent of the suffering of both mother and infant; a total of 13,248 babies were taken at birth or shortly afterwards between 2007 and 2014 (ONS figures). In 2008, 0.1% of live births entered 'the system' from a year's total of 672,809 live births. By 2013 this had risen to 0.3% from a year's total of 664,517 live births. Only 10% were ever returned.

Social workers take the next baby and then the next, in the performance of their duties as laid down by the guidelines. Here's the experience of one retired family court judge:

" The work of Family Courts for years has been removing the second, the third, the fourth child from the same mother. Not infrequently, the sixth, the seventh, the eighth – In one case I've removed the 14th. And I know of two judges that have removed the 15th children from the same mother."

The youngest recorded age of the mother was 14 years. Babies often are removed quickly, at birth or soon after.

"In 42% of cases, infants were subject to care proceedings within a month of their birth. In 70% within the first year. The mother has little time to turn her life around." Dr Karen Broadhurst.

The average time between court appearances is 17 months. "Repeat care cases cause enormous suffering." (Mike Shaw, Family Drug and Alcohol Court).

The idealised world of the advanced western economies is very far from any ideal. Drugs and alcohol are the scourge of the west, another form (deadly in this case) of pseudo bounty. Rewards still accrue to the most inept or unfortunate members of society, leading to the worst crimes and types of bestially one can imagine. Nothing our earliest ancestors could have perpetrated comes close to the degraded behaviour of some of the citizens of our most 'advanced' nations. One recent case of an infant being battered involved the male partner of a mother whose baby for her own safety had to be taken into care, probably to save her life – she was 3 1/2 months old.

The pattern of behaviour is familiar and predictable; the state, with a sense of benign patriarchy, invests itself with the collective sense of power from a male caucus, assuming the role of supreme father, leader and controller. This powerful male group are the inheritors of the skills and temperament of the ancient hunter-killers.

The physical strength once used to compete against other males for the opportunity to mate with an 'ideal' female is distilled to become a more powerful influence. Drugs act to conjure legendary power; users become imagined 'Gods'. No wonder male drug-dealing bands in powerful groupings dominate urban landscapes cultures. No wonder marginalised women are preyed upon.

As drugs become symbols of power, and alcohol creates illusions of strength and indomitability, women and their young become victims. As the Family Drug and Alcohol Court is discovering, women are crying out to live a natural life. The men at both the top and bottom of society are desperate to keep control, as societies take on the proportions of states ungovernable.

Many wives are enslaved to abusive men. Worse, in many instances they themselves are born to abused mothers, children destined at an early age to become pregnant by a carbon copy of their father. Husbands in deprived communities meanwhile act out their lives according to a primitive human blueprint; any opportunity to resurrect the hunter-killer is what attracts the male. Instead of a fit, brave male in the valued role of a member of a skilled hunting group, the danger today is a subculture of crime offering every disadvantaged male a substitute to the impossible dream of an organised life; the primitive adrenaline rush, empowering him and conferring a pseudo-male identity.

For such a male, outlawed substances have a risky, outlawed taint. The user feels powerful because he has access to powerful substances. Many teenage males find this culture irresistible; access to dangerous, sought-after substances empowers him, reawakening a primitive sense of self. Alcohol or drugs provide the misty heaven of self-belief.

Some other source of self-belief, some chance to build self-confidence or sense of identity in a fractured, irretrievably isolated urban environment is hard to come by. Outranked by any State appointee, fathers are emasculated. The role of mother is categorised as one that is demeaning, by definition, dependent upon pre-natal care, medicine, scans, midwifery, nursing and aftercare. A culture of care and protection permeates every aspect of life, as if the State is responsible for the tiniest 'bump in the road' but, at the same time, is incapable of fixing it.

Following centuries of continuous patriarchy, the widely accepted doctrine was that a supreme Patriarch directed life's affairs, an 'all-seeing God'. The State now effortlessly assumes the role. The appointees of a hierarchy were already in place – a male hierarchy.

Later systems of government formulated themselves in a similar mould, as direct descendants of Puritanism or Catholicism. Certainly, Whigs and Tories can trace their ancestry directly to previous religious groups – they were at each other's throats in the 17th century, and still are.

Socialism in particular aspires towards a more patriarchal stance among competing ideologies with unintentional consequences. In the desire to maximise social provision and justice (thus both to justify as well as demonstrate the efficacy of policy), the largest possible number of people must benefit from the patriarchal care and largesse. If this outcome is successful, fewer people require the services provided, and through this diminished need, logic would suggest the regime itself becomes superfluous. But no.

This self-limiting effect might reduce the influence of patriarchal power. By restricting or rationing services, through lack of finance, for example, the 'demand' for benefits continues uninterrupted, thereby justifying the ideology. It is a sobering fact that many socialist councils are inevitably 'safe labour seats', often for generations of successive socialist candidates, by means of this ideological 'sleight of hand'.

In these environments, in every 'advanced' city, there is ample evidence that men and women are captive to artificial or perverted roles; that is, this is not 'what nature intended'. Though secured over millennia of human evolution, their respective identities have been usurped by whatever the current 'State' has managed coercively to organise; highly capitalistic at one extreme and repressively socialistic at the other.

In every city, enclaves exist consisting of a spectrum of criminals, petty criminals, abusers, warring gangs and exploiters. These may control the local trade in drugs or the high finance of entire nations, even governments, commerce or trade unions; some police forces fit the bill. They have in common the fact that males are reacting to primitive drives that today have no means of 'legitimate' social expression. The visceral drive of animal power has not diminished over 5 million years.

Behind social organisation is recall. It attempts to resolve the conflict between consciousness in the human and the unflinching implacability of 'nature', 'natural phenomena', 'the natural world'. In the latter, there is no 'good' or 'evil'. These reference points, so ardently and endlessly sought by the human, are the work of consciousness, with no *raison d'être* in nature, no place.

So humans are a paradox, as is the power to reason; we cannot resolve the conflict between our 'natural' inclinations as men and women with the 'need' to reason our way into an organised condition, or out of a disorganised one. Paradoxically, we search amid the chaos of nature for an ordered world that fulfils all these

163

appetites and at the same time accommodates human organisational ideology.

The truth is that recall is the result of mutation, a natural 'flaw'. From a its first appearance, recall, as Darwin might have said 'set the evolutionary cat among the pigeons'.

In the simple but chaotic landscape of nature, genes present in offspring will, for females, express for (engender) the ability to survive effectively in a particular niche, raise young, display those attributes that attract a number of suitable males with one of whom she will accept as a mating partner. No organisation needed.

For males, genes present in male offspring will reflect those of the parent;   physical ability to compete with others exploiting the niche, for space, food and water, for sufficient range, but above all for the best female of his species within range, ensuring his genes 'wins' the 'fair lady.

In contrast with our civilised world, this seems a simple formula. It is not. The genetic code for any creature is vast; we have identified only a small percentage of the working components of our own DNA. Most of our DNA is a mystery, though what appears to be indecipherable may relate to mental architecture - how recall structures neurones.

This is likely to be as complex as the mental niche itself and the way we have constructed it; the language, logic, illogic, the illusion of past and present, indeed everything our artificial niche contains, including a large element we will never be able to decipher. We are captives in our niche, just as are other species in theirs, except ours is of such an extent that they, too, are now captive in ours - human influence is now universal.

Contrasting strongly with the modern notion of 'liberation' is the fact of how imprisoned we are; in our sexual psyche (primitive), in our society (patriarchal), in illusory structures that strive to impose 'order' (science), in the mental niche itself.

Now defunct religions once insisted upon female chastity. Women were – and in some cases still are – bound to ideas of purity and cleanliness. With good reason. The Virgin Mary was purity herself, so pure, that the Christ-child carried no conflicting 'baggage' in the way of someone else's genes. No male genes or heritage would make claim upon him subsequently. So virgins are pure and therefore ideal mates, unlikely to already be pregnant by another male. We can see again the inherent fear of the males for the inherently mysterious workings of the female. If she is 'un-choosy' then she may be a bad breeder, or already pregnant by an unknown male. Her final choice, if other choices have already been made, doesn't confer any special qualities to the 'last in line'. The male, instinctively, still wants to be 'first in line'.

In the modern State, these complexities born of our natural selves persist, in films, novel, songs. Every decent plot combines these skilfully. Blockbusters cram in every possible ingredient; sex, lust, aggression, war, heroism, betrayal (including sexual 'betrayal'), frailty, brutality, nobility, love, dominance, defeat, resilience; the very human characteristics the State would rather eliminate from their already unruly populace. Art and leisure, films and books are the safety valves for society. Yet as human characteristics become blurred in the individual, the greater is the need for their representation in art, music, film and literature. Even journalism is roped in to supply that need; where once sober reporting of fact was its singular role, sensation reigns.

Entertainment is now such an industrial behemoth that revenues rival that of nation states. Now the need for human expression contests with patriarchal dominance, within the territory that is the ancient battle-ground; human sexual psyches battle to stay afloat in a sea of social complexity, organisation and control. Not even the rigours of enforced communism survived the evolutionary power of sexual reproduction. The collective ideal eventually bowed to the sheer persistent illogic of the human and their intimate, personal drives.

CHAPTER 10

The Scandal of School

Humans are trapped in designated sexual psyches assigned to us more than three million years ago. Our developed modern society doesn't help; our natural inclinations, unlike any other creature on Earth are forcibly suppressed, first by parenting and then by schooling. Preparing for the complexities of modern life needs care and attention; a wider organisation of society has supplanted individual action and instinct.

As an example, our schools, instead of calm, focussed places of learning are cauldrons of trouble, a concentrated atmosphere steeped in gusts of pheromones and awash with bubbling hormones.

At first sight, school for a young person may be both exciting and daunting. Yet this environment is fraught with difficulties; degrees of pain, suffering, failure humiliation and derision that can be vicious and hurtful. Young people can be mentally (even physically) 'scarred for life' by their experiences.

With hormone levels churning riotously, emotional highs and lows are scary fairground rides for some. Others find school liberating and positive. Is this really the time and place to educate the young? Or is this 'socialising' environment in place to quell imagination and academic skill, rather than encourage their development?

Many teachers report that teaching is sometimes a low priority; keeping order is half the battle. Why are children, captive already in their different psyches, forced into a maelstrom of encounters, like actors on a stage with no script? Boys swagger and compete in tests of threat, bravura or minor physical violence, endless verbal jousts; the primitive male is there, just under the skin.

A heady mix of young male and female psyches presages sudden and sweeping changes that only detract from the formal education of the young.

The question of hemlines must be raised again; girls leave home having passed muster, only to wind up their skirts at the waist in a competition with other girls, daring each other to be the more risqué (with no end in sight).

There is very little reason for males and females to be educated together. By the nature of human evolution, it might be found that young males are best at practical tasks; encouraging the application linear logic, formulating clear objectives and challenges of problem solving especially those involving the use of tools achieving worthwhile outcomes.

Females are perhaps better at subtle discrimination, relishing varied ideas that need evaluation, collecting facts and arranging them appropriately, to be stored for later use.

'Education' seems intent on methods designed to force young adults to learn together whereas, arguably, better moods, attitudes and self-confidence are likely to result from separate education. Private schools seem to support this. Is the State intent on introducing some arcane ideology into education? Certainly there is a case to be made for more respect, confidence and better learning if schools and universities were devoid of this super-charged and wholly artificial atmosphere; encounters between so many young of the opposite sex is wholly unnatural. One school may comprise several hundred pupils. Few other species exhibit such odd behaviour; young, fast-maturing individuals in effect are inevitably making choices based on an as-yet under-developed urge to mate. They are literally 'groping their way'.

From the age of 8 or 9, up to becoming adults at 17 or 18 or beyond, those in education may be meeting a far larger concentration of people at close hand than they ever will in adult life. No other pubescent animal is 'thrown in at the deep end' and expected to behave maturely. Other species have parents always present, excepting perhaps penguins.

Collective education is an experiment by well-meaning but politically motivated adults. Schools of two to four hundred pupils in a state of emotional and hormonal turmoil are a misguided concept, ignoring the role of parents. Mixed-sex education is a bold but puritanical model, caring nothing for fragile emotions and egos. Even mature, experienced adults would find the kind of sensations and feelings of a new school hard to cope with.

The lack of a calm environment but, instead, 'one-scaled' up to meet the economic and political demands of State systems fulfilling the worthy notions of the 1930's, means that both masculinity at its best and femininity at its best are usurped by dubious mix of social dogma, facile ideology and 'economy of scale' at its very worst.

It's a 'fact of life', as they say, that there would be no men without women and no women without men. Yet today it takes humans perhaps twenty years to reach a point of sexual and emotional maturity. It takes longer with every passing decade as the sheer informational load expands. From this obvious, entirely natural state of affairs, there is no need to find a political or religious framework. Instead, let's make use of the male propensity for male linear logic to educate our men-folk. Let the mental versatility and sensory intuition of women educate our young women. Let them cross over whenever they want, from one side to the other, but let male logic and female intuition drive the core subjects, not patriarchal ambition for control.

Later, in real life, women and men learn to avoid the folly of competing with one another. They know instinctively that each has a separate psyche. The unspoken propaganda of mass education – it's a 'given' that all people are equal, and must be treated as the same – is behind them, thank heaven, to be replaced by the single and signal truth that education fails to teach; there is no equality except in law, the propaganda was wrong.

There is a sense that modern society is tending (via this methodology) to perpetuate childish attitudes and behaviour. Adults now indulge in child worship as birth rates decline; children are more precious. Adults delay starting a family. The 'valued child', wears Gucci shoes and mothers dress female children in 'glitter and heels'. Noticeable, too, is that tendency of adults to speak with infantile pronunciation, as if to defuse what they say, to avoid contention, perhaps even to avoid sounding adult. Perhaps to perpetuate the comfortable notion that they don't have to mature, no frightening independence is required of them. An independent adult might fail, children cannot. It's frightening to escape from State control, the ultimate adult, the almighty patriarch, so let us succumb; defer to the ultimate adult.

It may be that, as the State pays such dutiful attention to our young, mothers abrogate responsibilities, imitate the gestures, language and type of behaviour that brought such rewards when they themselves were children. Fathers adopt the mannerisms condoned and celebrated so assiduously; it gets them out of the challenges to be 'faced' by the male in a severely controlled world; children are no longer "seen and not heard", they are with us forever.

Meanwhile, children well-cared for in advanced countries are neglected and abandoned in others. Many of the world's children never go to school. No Gucci shoes for them, except in the unlikely event that the bi-monthly discards from one economy charitably end up in another.

Education tends to be a cast-off as well. In poorer or less-developed regions, education is likely to be simple and practical affair, repeating custom and culture and the knowledge useful for day-to-day survival. The cast-off kind of education is the rudimentary and discarded or outmoded beliefs of some other, better-off nation. Charitable donations of education reflect the principles and values – by-and-large – of the educating donor.

Unfortunate examples from history reveal how the 'Western' conviction of the benefits of education comes to mean the 'benefits of Western' education. In one notorious example, the once unique and self- contained cultures of the Pacific islands, including Hawaii and its archipelago, fell prey to Christian missionaries, their religion and to their illnesses. Hawaiian beliefs required an honoured individual to have the flesh 'flensed' from the bone. The bones were placed, after burning the flesh, in sacred grottos or caves, a site of reverence, family honour and tradition.

Christian missionaries, particularly from America, condemned this practice as ungodly. Quite why interring the entire corpse underground, to allow it to rot and moulder away unseen, could be considered more sacred is unclear.

That essential ambition of 'missions' to convert 'savages' may derive (especially in men) from the aggression and logic inherent in patriarchal systems. As with Puritanism, the conviction of rightness *is* the religion; the act of vigorously, sometime cruelly, enforcing belief has the happy effect of removing any lingering doubts in the mind of the perpetrator. The conviction of being right is also a distinctly male characteristic. We must add in the less-than-divine greed for new territory, wealth and resources.

This leads us to a strange and sad paradox; those who adhere, practice and proselytize most fervently, whose personal mission is to become exemplary models of their 'faith', are those most likely to abuse and humiliate their fellow members of the human race. This applies both to children and to those of an unfamiliar culture. It is no surprise that missionaries favoured the idea of indigenous people being 'childlike'. Mankind – sorry, humankind – is known collectively as the 'children of god'.

And so, another paradox; the scandal is that school is a territory that appeals to those whose mission in life is to dominate, as well as subtly to convey their extra-curricular beliefs and opinions to innocent and receptive minds; a fertile breeding ground, like the mission station, for political indoctrination. The church school is a junior 'mission station' for early indoctrination. The

State school is also a mission station for political indoctrination, witness the National Union of Teachers. In an expanding mental niche, both the quantity and increasing diversity of information present an ever-greater challenge to teach with expertise and moral judgement.

The teacher's lot is not a happy one, but instinct and natural ability are trumped by linear logic harnessed to an obsession with endless measurement. The rigours and demands of the classroom are unfair to both teachers and pupils. No attempt is made to encourage qualities that don't lend themselves to measurement; patience, empathy, insight, intuition, charm, personality and humour go unrewarded in both the student and the teacher. The teaching environment is wrong; the schools are too big, curriculum too sketchy and piecemeal, the underlying philosophy archaic and misguided.

Patriarchy invades education through its administration, its scale and objectives; it seeks to cram children and students with a smattering of knowledge, a 'representative sample', a shallow 'cross-section' of the mental niche. How real society works is passed over; how every shop has to keep accounts, how every parent has to pay tax, how electricity reaches each home, who runs the patriarchy. This messy view of education is typically male and typically patriarchal; that States are somehow better equipped to decide the fate of children than are parents. Parental roles diminish, sometimes to the point that parents are still childlike despite having children of their own, their adult roles suborned. Feeble mental preparation is apparent even at graduate level and passed on to the children of graduates. Patriarchy rules supreme and is supremely ignorant, sufficient unto itself. Humanity and individuality are the enemies of State. Female personality encapsulates those indefinable qualities, qualities so difficult to comprehend, define and control. Patriarchy takes a dim view of women.

Worse still, the mental niche expands exponentially, encouraged by the state, blithely unaware even of its existence. Modern populations have to adsorb unprecedented amounts and variety of information. This flood of stimuli and data means the moral and intellectual capabilities of individual humans will begin to fail if they are required to absorb it all. The expectation that individuals can retain forever-increasing volumes of data is, to say the least, unrealistic.

Unlike a computer, humans may fail to adjust to the exponential growth of the mental niche. We are physically primitive. We cope with complexity only by our *collective* intelligence. Separated from the collective we revert to individual human behaviour, the different psyches of male and female. We need to retain these separate roles for they are of great and lasting value in the game of survival. They have stood the test of time. The 'informational blender', a recent phenomena comprising the expanding mental niche and the expanding patriarchy may well obliterate, or at least lobotomise, humanity.

CHAPTER 11

'Yours, worried'. Origins of Female Anxiety

Every newspaper or magazine has an 'Agony Aunt'. Most of 'her' correspondents being female. Anxiety apparently afflicts women more than men, though perhaps women are more likely overtly to express anxiety, while men see this as a sign of weakness. Equally plausible is that there is a long-standing cause, hitherto overlooked.

The female presentiment is something we've encountered before. It is perhaps inherent in her archetypal psyche, encompassing a broader spectrum of sensory attributes. Our shorthand version is that males employ linear logic while females have sensory intuition. Recall is, perhaps, more highly developed in her. Maybe it enables more detailed assessment and subsequent evaluation, the ability to build a more detailed memory repertoire from which is derived deeper insight.

This is not the mental toolbox of the male, but completely different. Instead, the female has acquired a risk/value assessment capability that would be, in the male, entirely counterproductive. The repertoire evolution has bequeathed the female has become an innate faculty used as a complex reference system. Based more in emotion than rationale, not wholly reliant upon conscious recall, it allows the female to assess more precisely, more intuitively, the flow of events, the character of people, the pertinence of objects or circumstances, their relevance and value.

The repertoire behind sensory intuition must be a phenomenal library, literally a library of phenomena and effective responses. It must consist of an extraordinary amount of data. There is no such thing as the 'empty-headed' women, so beloved of (hopefully, now defunct) male critics. The quality of data must also be measured and stored, assessed and compared; each new experience logged with a 'value' signature, to aid future reference and recall. This is hard for the male to analyse, quantify and describe, let alone understand. Suffice to say that women will 'get the picture', in a way wholly familiar to them but difficult to express. This is the nature of sensory intuition. Which is why females can be prey to anxiety: this level of intuition has its drawbacks, principally the presentiment that events with so many possible outcomes – she can imagine them all – have the potential to wreak havoc.

The archetype is probably consistent with the period when males began to develop co-operative hunting techniques and women became co-operative 'garners'. This more specific term (I hope) encompasses more than just 'gatherers' in that it includes the qualitative element mentioned above. This the ability to collect diverse edible fruits, roots, leaves, etc., much in the same way that forest-, desert- or savannah- dwelling people do today.There's no point in 'gathering' alone; garnering means (I hope) carefully judging how much to take; what will satisfy immediate needs, how much needs to be left for later, next month, next year? If all the eggs are taken from the nest there will be none next year.

Considerable expertise is also required to recognise non-food material, may it have medicinal properties or uses as the base for fibre, fabric, utensils, or ritual. But, while the garners work, their protective body guard is elsewhere. Does female anxiety stem from this? We might conjecture that the four million year-old evidence of flint workings, suggesting widespread use of arrow and spear heads, axes and knives, coincides with specialisations – the division of labour – with the dangers inherent in garnering at least as risky as the hunt. For women, the world changed forever.

Did anxiety first beset the women of the clan when they set out accompanied only by children and elders to search for the essentials of life? Their mission was to harvest food from sources they knew would be naturally replenished, but their natural protectors were otherwise engaged. The dilemma is further heightened as their maternal role is made more dangerous.

The sexual psyche of the female was born long before the scenario outlined above, in an age when femininity most likely enshrined the sensory intuition necessary for nomadic life in the wild. It is likely that men were more like women at that time. The advent of the arms race ushered in during the Palaeolithic changed humans

178

forever; the new human was now an aggressor, at war with nature. Division of labour, creating two separate and specialist psyches, drove on that success and humans would never be the same again.

But, in women, vestiges of the archetype persist, found in the echoes of her sensory and intuitive abilities. In men, these qualities may have been supplanted in the Neolithic period by a 'fight _not_ flight' response that proved sufficiently successful that evolution favoured this response over 'flight _or_ fight'.

None of this can be stated as a definite outcome, but oestrogen (like testosterone), one of the key bio-chemical components in 'signalling cascades', found in both vertebrates and insects. Thus, estrogenic sex hormones have an ancient evolutionary history pre-dating humans by millions of years. So, successful flight responses select for species survival.

Another vital biochemical response is the action of corticotrophin-releasing hormone (CRH) secreted by a cluster of cells (the par ventricular nucleus of the hypothalamus) in response to stress.

In plain language, female sexual psyche may be more ancient in origin than is its male counterpart. Stress and anxiety disorders occurring from puberty to age fifty in women are twice the level found in men. But CRH can't be modified or somehow reduced in concentration, its effect ameliorated by administering some palliative; CRH is part of the vital mechanism in the placenta, acting as a marker to determine length of gestation, determining also when parturition will occur, rapid increased concentration acting as a 'trigger' for birth. While the female skeleton has adapted to life with an upright gait, there is nothing 'modern' about CRH; women are captive to their biochemistry and this may account for a predisposition to anxiety.

Modern life is at odds with prehistory. Women would, we can conjecture, feel perfectly 'at home' with life before the advent of tools or the appearance of armed and lethal bands of hunters. Smaller family units, the occasional encounter with another tribe or group, a slower pace of life, browsing on familiar foods (avoiding competition with superior predators) and with an innate familiarity with the ways of the world, marked a peaceful time, lasting aeons.

As with other inhabitants of the world, recall was superfluous; instinct and memory prompted by simple cues from the weather and the environment, stirrings of internal biochemistry and hormonal responses formed a basic rhythm of life. We can still see these patterns in free ranging primates and other animals. This is what we once were like.

The 'signals' or 'stimuli' from the physical environment prompted balanced individual responses shaped by evolution. Our physical niche is gone forever. In its place the expanding mental niche is growing daily and with it every possible combination of events. Now we struggle to deal with the constant 'signal cascade' from an abstract environment of ever-increasing scale. We need rafts of mechanical aids to filter and order them. Without them, we are incapable of dealing with life at this degree of complexity.

The female psyche is of more than prehistoric origin; it is the result of hundreds of millions of years of evolution. These specialised processes are of such bewildering complexity that they are as yet not fully understood. They are obviously robust (they've stood the test of time), but as men changed – those less 'brave' less favoured for breeding – a simpler profile emerged; genes expressing for linear logic and encouraging the survival of those men more adept at weaponry and so favouring a 'simpler' male psyche. Its success is apparent in the mental niche and its rampant growth. More complex is the nuanced process of sensory stimuli and the responses they evoke. Embedded in more complex biology (involving more than just the brain), and characterised by subtler outcomes, feedback to the parasympathetic nervous system does not admit of exploration.

It may turn out that linear logic is a barrier to the description of sensory intuition. Certainly, it's hard to measure. For the male, it may not even exist; if it has no logical explanation, it has no existence. We have to remember here that the existence of wombats, for example, is beyond question. The wombat, however, displays no capacity for exercising logic. In the wombat's view, wombats don't exist.

Sadly, perhaps, we are not wombats. Women diverged from men, or perhaps men diverged from women. For women today the anticipation of future outcomes in a world crammed full of possible threats. To try to anticipate every eventuality may be stimulating or may be a daily nightmare. As a wealth of possibilities grows without any apparent limit, the process of "*intracellular signalling cascade*" or, subsequently, "*signal transduction*", takes on ominous meaning; our bodies and our minds do fail – not catastrophically but gradually and imperceptibly.

In women, the 'flight or fight' mechanism is, given the fact that they probably have not followed the male evolutionary route, more 'sensitive' or 'highly developed'. Partly as a result of oestrogen and progesterone means the mechanism is activated more readily (we all know that feeling) and stays activated longer (we know that one, too). The neurotransmitter serotonin also plays a role in response to anxiety and stress. Unfortunately, (wouldn't you know it?) women don't process serotonin as quickly as men. Women, as we've seen, are more sensitive to corticotrophin releasing factor (of course) which organises stress responses, which is why twice as many women suffer from stress-related disorder. But, this is *not* a fault in women; we have seen that these bio-chemical processes are tried and tested, proven over millions of years to favour survival. It is the environment that is wrong, *not* women's biochemistry, attitude, her personality or her sex.

Agony 'Aunts' need a crash course in evolutionary history, as do men who complain of women's 'worrying over nothing', 'fretting', 'fussing', being 'neurotic'. The conditions are real. Physical pressure from the environment causes stress; noise, overcrowding, competition for space, lack of sunlight (low vitamin D'), poor food (ditto; other vitamins), pollution from chemicals, exhaust fumes, electrical radiation, house mites, fungal spores, electronic radiation (TV and internet), the list is endless.

Zookeepers go to great lengths to ensure their exhibits are treated 'humanely'. This is a supreme irony as, increasingly, humans inhabit an environment unfit for animals.It is also ironic, then, that when women fall prey to anxiety and stress their condition is systematically and consciously downgraded to just 'worries'; an indefinable (and therefore irrelevant) condition. The condition, like a canary in a cage, is serious and needs addressing urgently.

Yet only a century ago women were deemed content if they had a home and children and a caring, responsible husband as 'breadwinner. Female skills, intelligence and sensory intuition as old as the hills reappeared given the slightest chance, to the great chagrin of champions of patriarchal society; education for women was frowned upon at best, at worst forbidden. To be educated and vote, to own property and sue for divorce were socially cataclysmic.

Thus released from tyranny, did women feel free? Still we hear of glass ceilings and unfair treatment in every sphere of endeavour. Again, it is the environment that is at fault; organised along historical lines that suit the patriarchal dictates of males, how can it change? Bizarrely, women are required, as mentioned before, to, follow male precepts and attitudes, to 'ape' men

A society based on matriarchal values is not enough it seems; without linear logic, women have no organisational agenda flowing naturally from their sexual psyche, nor do they have an over-arching impulse to control and dominate. Theirs is not an organisational will, but a nature that reflects nature, not one beset by implacable logic. The society that follows a natural course, rather than an artificial one, chimes with women; the basic rigidity of our organisational structure has to evolve to achieve some redress, for nature and for humans, but especially for women; 'soft' governance.

This is a tall order. Calling into question the mechanisms of control, power and dominance directly threatens the male hierarchy. It is unlikely that any male, equipped with linear logic, descended from a long line of phenomenally successful hunter-killers, could ever be convinced of the need to change the way we co-exist with nature. One day, women perhaps might take the lead in this.

CHAPTER 12

Symbolic sex

The changing mental niche has sidelined women; mechanisms devised by men underpin the modern world. The inventors, owners and operators of much of the devices that make the world easier to manage – for both men and women – are not of her design, nor of her choosing. Linear logic has such a hold upon the male consciousness that it has profound expression in powerful devices.

Men love guns; women hate them. Men drive cars; they drive fast and dangerously, given the chance. Risk, speed, noise, action; these are the ingredients that men look for in life. The pressure is on for women to join in, to subscribe to the life styled for budding patriarchs.

Many do, encouraged by a 'sisterhood', mistaken in the belief that sexual equality is nothing to do with law, but all to do with power. Thus, competing with men on equal terms brings them the power necessary to gain acceptance, to succeed. This is a false hope; the status of women declines in direct relation to the degree to which they reject femininity and adopt male politics and male socio-sexual attitudes. Fearful males abuse or denigrate the sexual psyche of women. They certainly don't have innate respect for women and from an early age adsorb

the one-dimensional view, the male response to nature –
to dominate, control, defeat. And it gets worse.

First encounters for the young are now rarely with real
people but, instead, with symbolic representations of the
sexes. Storybooks peddle unreal stories of stereotypes.
Cartoons, TV shows and comic characters reflect
idealised or laughable versions of future mates.
Symbolic sex takes the form of 'soft porn' even for pre-
teenage youngsters; girls with idealised shapes, pouting
lips, blond flick-ups meeting bragging boys in check
shirts performing implausibly heroic tasks.

Appropriate for childhood perhaps, except the rot has
already set in: few females are role models or leaders.
Boys take the lead, are the heroes, save the girls, solve
the problem and catch the thief. Stories for young girls
teach nothing about the respect to which they are
entitled. Productions aimed at the young already bear the
taint of commercial propaganda.

Sexuality becomes increasingly symbolic; how many
couples form relationships purely by personal choices?
Girl-meets-boy is increasingly a rare event. Instead, the
ground is pre-prepared by skilled branding professionals
with images loaded with unrealistic promises; girls are
impossibly beautiful and young men dashing and
handsome even as they approach puberty.

Women's roles are trivialised by an increasing clamour
for glamour; only this lipstick or that shampoo will
confer true feminine attractiveness. Competition between

cosmetic brands dominates in a world in which media controls the selling space and manipulates the levers of cultural influence; mothers and fathers can't compete.

Single women face a dilemma. After education comes independence, together with the expectation that an adult individual is self-contained and self- supporting; wealth and independence are the 'offer' society favours, with a strong dose of glamour and sex appeal. Individual choice is the message evoked by advertising: this product is 'just for you; our seductive message is carefully crafted to set you apart' (from the mass-market). Consumers embrace the delusion that engenders a haughty disdain for anything that smacks of compromise.

The message is successful: the elaborate courtship (though, in this instance, courtesy of commerce) fulfils the evolutionary 'purpose' essential to the continuing survival of any species; mating will produce more 'customers'. But let's hope they break up soon; two households are better than one.

Hence, entering marriage or a partnership, 'doing what society expects' leads to contradictory and confusing outcomes. Having been 'sold' the image of women as powerful, individual, influential citizens, 'equal' in every aspect, our growing young woman encounters disappointment in marriage or partnership. Confronted by legal conundrums and social prejudices, the cynical truth is that the powerful economic forces at the core of modern life are not conducive to family life. Much better is a continuing round of product purchases.

Women might choose (perhaps unconsciously) to continue life as a 'young consumer'. Self –image will confine them to that role as long as it continues believable. The attractions of adulthood (a respectable married woman is a part of long-lost history and deeply unfashionable) consist in disposable income as a mark of personal status. It is also the key driver of commerce; why encourage thrifty couples when encouraging the continuance of two separate disposable incomes is more profitable? Small economies might characterise a careful housekeeper; the commercial world abandons them to their fate – oblivion is for those who fail to spend.

Once characteristic of male status, possessing a car is now de rigueur for a young woman; it confers mobility, freedom and safety (yes, women are still generally safe drivers than men). Travel also adds to status, often measured in lists of destinations, the more far-flung, the better.

In political and economic terms, in this way maturing women are encouraged subtly to 'masculinise' as the appeal of 'young-teen' persona fades or becomes obviously absent. Cosmetics, after all, have their limitations. In a modern, idealised role, with more discriminating taste and greater disposable income, our sophisticated young woman takes on her male competitors. Along the way, reaching age 30, she may seriously consider marriage, for 30 years is now the average marrying age among women (in the UK).

This being a convenient though arbitrary 'marker' for the sake of our argument, women may not have children after marriage or they may already have children. Here we have to consider the time span between puberty and 'marriage', leaving aside the average age of women at the time of birth of their first child. We need these markers to examine the change – in this case quite rapid – in behaviour as determined by, or affecting, sexual psyche.

There is an ever-increasing time span from the end of childhood to childbearing age. Childbearing is the evolutionary raison d'être for humankind, as for other sexual creatures. The human female now has a lot more time than in any preceding epoch to savour (or otherwise) the possibilities of stepping outside their sexual psyche; perhaps to decide to forego motherhood altogether, in favour of an evolutionary 'gap year'.

That gap is widening. Onset of puberty is now occurring at 12.9 years in Europe* roughly equivalent to girls in India in AD 300. In the Middle Ages in Europe the age was 14, while in ancient Rome some girls were bearing children at age 11 or 12. In Europe at the start of the 19th century, 17 years was the average age but this declined at about 4 months per decade as societies changed from agrarian, rural groups to industrialised urban communities.

The changes are growing rapidly with every passing year in the modern age; obesity and endocrine disruption through pollution levels over the last 50-60 years has an effect - modern life has been an unkind friend to women from one perspective, or a liberating experience. The birth rate (per woman) in the UK is now around 1.9 children, a 12% rise between years 2003-2012. This is mainly the result of immigration, new citizens not subject to the same demographic trends as the indigenous population. British-born women have fewer children, and finances, house prices and concerns over welfare and employment delay child bearing. (*ONS*)

Older women often seek help in conceiving a child – 'older' here means a median age of 33.5 years (1991) rising to 35.11 years (2010). Those seeking IVF treatment also include males, outnumbering women by a healthy margin.

Perhaps men feel it is important (to their status and identity) if a baby – preferably male – is born, his wife then (incidentally) becomes irrevocably tied to family life, her freedom, mobility and influence diminished?

The records show* that the number of children born following IVF treatment rose between 1991and 2009 from 2,162 to 15,316. Techniques improve, along with information and confidence, though declining health factors may impinge. *Human Fertilisation and Embryology Authority*

One conclusion must be that the gap is widening between onset of puberty and pregnancy. This is a matter of choice, not so much a matter of chance, as it was less than a century ago, and throughout female history. There are simple choices made using 'the morning-after pill' and other forms of elective contraception.

Let us revisit the Pyrenees and the (rather down-to-earth) analogy of the stallion and the mare. The stallion has his way and there are no consequences. His only nagging worry is that a male interloper may challenge his dominance of the herd and the threat of predatory carnivores. The mare, saddled with pregnancy, birth, a suckling infant and the stress in guarding a virtually defenceless foal; ideal meat for a roving predator. Despite the unprecedented speed and scope of our evolutionary advance, the human psyche is still subject to primitive imperatives formed in the Pleistocene and Palaeolithic. IVF is not only a revolutionary advance, but also one that is evolutionary.

Our modern young lady is now spoiled for choice. The 'morning-after pill' and other contraceptives are direct evidence supporting the theory of Neoteric Evolution[1]; that humans have diverged from the classical evolutionary pathways that apply to all other species.

A long and artificially induced passage of time now separates puberty from adulthood in the female human. Reproduction is no longer a prime role, until she chooses. Like a student's 'gap year', this hiatus is contrary to nature's precedents; 30 years is a long time for any female of any species not to bear young, spanning that period during which it is most likely to occur.

---

[1]Neoteric Evolution. See Appendix.

This artificially imposed hiatus following from personal choice is without parallel in nature. A force for survival at any cost fundamentally underwrites natural selection. The human female is the singular example questioning this assumption. Leaving aside disabilities or unusual circumstances that might prevent breeding, reproduction is the existential driver of life.

It is in this context that the notion of 'symbolic sex' needs consideration; humans engineer the 'natural course of events' or modify them to ensure the sex act *fails*. Paradoxically, all kinds of benefits accrue; society can better employ women (and men) in 'more valued roles'. In performing social and economic functions specifically 'selected' by a well-organised and reactive society, the 30-year gap reduces the size of the population that has demands upon its resources. It increases opportunities to groom individuals until ready to occupy diverse roles as they become available.

The last thing a complex, technologically expanding, modern society wants is a booming baby population. Paradoxically, many Western societies suffer population deficit, made up for by immigration. The subtle difference, in this need for new babies (artificially supplied by burgeoning demand for IVF and other techniques), is that they be born to parents preferably already educated, conversant with and reflecting established cultural values and characteristics. Immigrants often fill 'low-value' roles, easily mechanised or dispensed with given the proliferation of 'labour saving' devices and solutions.

Births to immigrant families make up for lost births elsewhere. It takes time to adjust to educational, social and economic factors cemented into the 'new' demographic structures immigrants encounter. Meanwhile, the mental niche is continuously expanding, and at an exponential rate. Births to indigenous families may decline, but requirements 'from the niche' are for a more specialised caste. There may be more to transferring from one niche to a more complex one than the acquisition of a mobile 'phone. Prejudice, fear, unrest; these are not the product of incipient dislike; they are concomitant with rapid evolutionary change.

Saddled with a sexual psyche from 6 million years ago, humans struggle with the conflicts presented by the new environment and the need to ensure they and their family survive. This natural impetus is deeply embedded, visceral and extends beyond self. 'Self' also depends on tribe, mates, offspring, parents. Excessively modern subtleties assume lofty ideals should shape humanity. The reality is that evolutionary forces pull like the tide. In an endless sea of humanity, the currents are increasing.

Our sexual psyches are now far from natural. Male and female identity is counterfeited, perhaps evolutionary pressure is responsible for diverting sexual instinct towards the artificial and symbolic. Real sex confronts directly with the artificial version. Control of reproduction, now sanctioned by societies within which it was once a mortal sin, is tacit acknowledgement of the population problem. Humans are breeding out of control,

193

the largest number born to those least likely to conform to a modern agenda: individual wealth, security, long life and continuous comfort.

These anomalies produce, in the West at least. phenomena such as symbolic sex. Yet not long ago - the year is 1660, the time of the Restoration - a prominent member of the Aristocracy, Anne Harrison, Lady Fanshaw, mother to 6 boys and 8 girls, records her own experiences, in stark contrast to our comfortable modern state;

"In the latter end of Summer I miscarried, when I was half gone with child, of three sons, in two hours, one after the other."

(The Diary of Anne Harrison, Lady Fanshaw.)

Her many miscarriages, taken with her live births of 14 children, would be horrific to a contemporary obstetrician. Such circumstances of childbirth are associated in the modern mind with poverty and poor education. Lady Fanshaw was a member of the restored King's entourage. Her husband was a long- serving public official and later a minister in the court of the restored monarchy.

In a later century, much closer to our own (1830-1909), Elizabeth Stewart had eight children living and two buried. Our ability to conceptualise birth control or induced pregnancy is very recent. Dealing with the death of an infant is now trauma of life-changing proportions.

That evolution is itself rapidly evolving becomes startlingly apparent. Just a century or less separates Elizabeth Stewart's offspring from the epoch in which symbolic sex emerges.

Today, it is a significant influence in (mainly) western society; introduced very early by parents, extending to mere infants symbolic sexual identity. 'Proto-humans' are best programmed early. The insistent need to produce offspring is one of the most potent drivers of adult behaviour. Nowadays, quality over-rides quantity. Just 1.9 of-an-infant concentrates the mind of new parents wonderfully, enshrining a little prince or princess as a rare and valued new human.

So, ideally, all of our children must be flawless, 'as perfect as possible' to fulfil new criteria, the status symbol for the micro-family, a concentrated focus of parental attitudes and aspirations. The pressure on a young child is enormous.

Symbolic sex is present throughout childhood and young adulthood. Images and objects for children are already sexually orientated and new technology delivers this directly to a consciousness already primed. The consciousness that our new individual occupies a special place means all this attention is simply 'normal'. That it is often delivered direct to the 'consumer' by internet carries with it (in its technological sophistication) an authority that neatly masquerades as 'parental consent'.

By the age of sexual maturity and childbearing age, the absence of real sex in favour of symbolic sex results, of course, in the absence of children. This may give rise to the need for women to adopt an alternative status, now that menarche, pregnancy, maternity, fecundity and lactation may be in the process of disappearing from real life. There's no room for any of this in a sexually symbolic world. They have become dirty words.

The solution is to for women capitulate to the constant pressure of the male patriarchal system; now that women are officially classified as intelligent beings, it further condescends that they may now go on to swell it ranks, not as childbearing women, with all its attendant 'inconveniences', but as 'symbolic' women. Commerce has the means to ensure a more-or-less-continuous ability to absorb them into a new form of captivity; lifelong coercion through endless reinforcement of the concept of sexually orientated symbolism.

We can 'act out' sex in various ways. If we're careful it won't disturb our lifestyle or career prospects. The images of sexy women proliferate as never before. Meanwhile, birth rates plummet.

That we are doing without proper sex and playing at it instead is borne out by the burgeoning commercial landscape. Supermarket isles for fashion items, make-up, hair-care, cosmetics, lingerie, translates into miles. This sales sector accounts for £26,000,000,000 per annum. None of it is aimed at real women – those who suffer period pains and consequences of fecundity that are now all-but-unmentionable.　　Most customers are 'teens', 'gappers' or 'greys'. These ugly terms hide and uglier reality; commerce trumps womanhood.

The encroachment of the female into male preserve of 'paid work' began with the exploitation of female skills and personality in revolutionary manufacturing techniques, particularly in weaving and machine minding. Her stamina, her sensory intuition and indefatigable ability to absorb and concentrate paid rich dividends. It did so, too, in the 'secretarial era'. The fallout from both wars accelerated the participation of women in the economy, though not in reproduction; adding to the workforce through childbirth is too slow for the burgeoning pseudo species. Women themselves became direct earners while automation and machines incepted by male innovators accelerated the scope and speed of human advance. Even today, childbirth is a lurking impediment to our prospective female employee seeking a 'start' somewhere. The prospect of pregnancy is loaded with negative associations; employers (not always male) have an innate prejudice. The time might come when a certificate of infertility, heavy lipstick and a smart line in male attire will force a crack in the 'glass ceiling'.

Symbolic sex is present in every form of media, from the ads on our jobs-seeker's mobile phone and laptop to cinema and TV screens. The message is consistent; glamour and beauty signifies fame and wealth. The cute hemline, lips gloss, sultry eyes and curled lashes bespeak of an individual turned out to slot right into the proscribed profile. She is available, but strictly not for motherhood .

This has become such a persistent proposition (anyone with a few bob now can 'pass muster'), that it is now of considerable discomfiture to the original role models of glamour, who must paradoxically now 'dress down' in search for anonymity.

Our young (-ish) male has a different but predictable strategy, old as the hills. The power he seeks is symbolic, taking the form of mechanical devices, tools, physical signs of male status; strong, robust, dependable, this character wants to harness the things that display and evoke the best sexual stereotype: ready to take on any danger, best at providing protection, strong muscles, eager brain and a good line in linear logic.

First on the list is a car or motor-bike. The linear logic translates directly into visceral, straight-line speed, the kind that killed James Dean. 'Bullet', the Steve McQeen film, might have been the favourite of his grandfather or father. Action at a distance and a psyche conjured directly from the hormonal to the real, encompasses cars, guns, planes, knives, football, rugby, anything indeed involving a point, being sharply defined, a defined target. These early individuals formed the basis for the group of hunter-killers that all other species feared.

Being part of a band (I mean here the armed and dangerous sort, not the feminised boy band) presents the notion of co-operative strength of men in general, thus adding to their appeal. Even the puniest member benefits from this co-operation. No wonder boys 'go courting' in groups. Overt energy in general and conspicuous behaviour in particular evoke essential masculinity; swimming in adrenaline and testosterone and, courtesy of the support from group behaviour, these men think themselves especially attractive and practically invincible.

In the same way in which fecundity is now a dirty word, so is aggression. But, just as the description above could apply to a roaming band of young male animals, aggression made humans dominant. These instincts, honed to perfection by time-honoured need, are now a preferred ingredient in the hunter-killer mix of sublimated motives driving, for example, high finance, Grand-Prix motor racing, football or rugby.

Pseudo-sexual satisfaction comes much more easily and with no noticeable after-effects to our maturing male; the maelstrom of symbolic sex carries only minor dangers, the worst being that he might not live up to flashy image, nor may his credit card.

The average frequency for male sexual encounters is not particularly daunting as a target; for our 18-20 years-old, 112 times a year is average, mostly with the same partner. Then the anticipated decline begins: 86 is the score for instances of real sex (though by now, perhaps, routine), for the 30-39 years old age group. 69 times annually for the 40-49 years old. After that, it 'tails off'.

Next time in the supermarket, check out the cosmetics counter for just sheer scale. There are increasing numbers of male product lines, and the trend is set to continue. As symbolic sex serves to bridge the ever-widening gap years, (from puberty to a stable relationship) so these individuals 'lose out' during this period: for both men and women, 'real' (i.e., productive) sexual encounters (with attendant emotions and behaviour) diminish in frequency. Something is being 'traded'. Do we 'trade-in' control of our sexual psyche to the State in exchange for the freedom to experience instant gratification. Is the State complicit?

This may be a consequence of niche expansion, whose underlying effects are not immediately apparent: most of human activity is now within the 'mental niche'. But, certainly, from the organisational point of view, people can't go on having sex, 'willy-nilly'. Economics and subtle social distractions divert humans from reproducing – as we've seen, we don't even produce 2 children per couple, but 1.9, and falling.

We've come a long way from what was once the archetype of fecundity, the Wittendorf 'Venus', a bulbous figure emphasising huge breasts and generous belly . Also a long way from the Aphrodite of Cnidos, the highly influential ideal of female beauty in classical times[2], beloved by Praxiteles in the shape and person of

---

2

Phrygne, the courtesan. With Phrygne and Aphrodite[3] the qualities of goddess and courtesan were deftly combined, and glamour was born.

What would they make of the captive wives now enjoying such a bounty of 'freedom' that they have become pseudo-male, choosing an extended youth and birthing when they choose, courtesy of IVF? What would male ancestors make of our dutiful sons of the patriarch state, enmeshed in almost unutterable complexity so that loss of their identity, too, may be irrevocable? Proud males? Proud females?

In a following chapter the possibility that our 'real' nature – the decline of, or even the obliteration of male and female sexual psyches– is examined in the context of the niche we've constructed. Is the niche itself enveloping and subsuming original humans and favouring the evolution of a new, sexually neutral species? Sex, as we shall presently conjecture, might disappear altogether.

---

[3]

CHAPTER 13

Hostages to Fortune

Males are susceptible to rich iconography; much of American culture, for example, is concerned with gun-toting cowboys. Perceived (even unconscious) erosion of the male identity puts gun control way beyond even Presidential ambition. Male Americans still enjoy the illusion of a vast landscape in which to roam, having an atavistic hunger for wide-open spaces that so closely resembles the surroundings of their African ancestors that the description 'African-American' takes on a new and sublime significance.

Their English cousins also struggle to find identity that fits modernity. Their native culture, with such an attenuated Out-of-Africa experience that no hints of primitive association survive, is reduced to gleaning a vestigial identity from Imperialism, the industrial working class, football, warfare (sometimes hardly distinguishable) James Bond, the Bourne Identity (good title) and other fictions.

So tenuous is the British male sense of identity that it is almost a veneer; the female sexual psyche is at least more ancient and genuine, while that of the male is painfully thin, specialised to Palaeolithic needs. Which is why, perhaps, 'male bonding' is the modern substitute for the long-lost hunter-killer. So poignant is their need that male identity closely correlates with role. Apart from just 'being male', outside war and football there is hardly a role for them.

Men must work to have a role. 'Husband' is not sufficiently challenging to be 'a role'. Remember, inside the male is a primitive set of drives involving sharp or heavy objects, action at a distance, a lot of pursuit, huge amounts of expended energy, memory for tools, weapons and how to make and deploy them, linear logic as a basic recipe – a first principle – on how to tackle the world.

Most of what he is or does is singularly inappropriate for raising a family. This role requires a complete subjugation of his abilities, a denial of the prime motivators that constitute his sexual psyche. He can be divorced or arrested (or both) if he gives in to any of the urges striving for an outlet. Men play or watch football, drive too fast, take apart mechanical devices and put them together again (or not). In a local street, their ideal 'shop window' is a car with the bonnet up. Their kind of film or video game is fast action and trying to blast something out of the way. Early onset mayhem arrives in infancy and continues through adult life until exhaustion sets in.

Attempts to civilise them are, by and large, successful but it does mean that men are in a deeper trap than their sexual partners; 'hostage husbands' is a carefully considered term. Modern life is a civilised cage constructed from hefty ancestral bonds, but specifically designed to both harness and contain male aggression and inventiveness. This latter combination is a toxic mix of ingenuity and risk-taking; think of the hydrogen bomb, or the latest anti-biotic.

The accumulated experience of 'Civilisation' is that the males have a boundless talent for leveraging recall and thereby to extend endlessly the mental niche.

At the same time, attempts to strengthen the necessary restraints on male behaviour have to increase in proportion to the power released by their inventiveness. Luckily, males in this way make themselves captive by their fascination in discovering new and powerful devices. Today this takes the form of designing devices that are as far removed from flint axes as it is possible to imagine. These are nevertheless at the 'cutting edge' of the inexorable advance of the pseudo species. Our 'tools' are so highly developed it is now likely that no one brain even understands them.

Quantum computing – now arriving in an experimental computing facility near you – involves not the use of barely visible miniaturised transistors but the subatomic structure of matter itself, simultaneously to carry out computing tasks of incalculable complexity. This is meant literally; an understanding of the process is confined to, and probably only expressible in, the quanta.

Our young man, or young woman, searching for a posting at the cutting edge, must have already given up the idea of 'settling down'. Disciplines are so rigorous that few individuals alone can now make a contribution. The pattern is for larger, specialised 'cadres' to work together in the sphere of high-level problem solving.

At best, life is a repetitive routine involving difficult juggling acts for young adults, captive even in pursuit of 'leisure'. 'Time' that is *really* 'free' is a rare commodity. Those males who identify strongly with their role are increasingly prone to depression, even suicide, if they lose their job or join the ranks of the long-term unemployed. Male suicide appears in official figures as 'more than' three times that of females. In fact, it is 3.725 times, or nearer *four* times that of their female counterparts.

Of total suicides in 1981, 63% were male, but by 2013 this figure had climbed to 78%. Suicide is the leading cause of death in men aged 20-34 years.[4]

The scarring effect of unemployment resulting in subsequent suicide among males is known to professionals but not to the wider world.[5] It turns out – though they wouldn't want you to know –that males feel hostage to a role to which many are not suited. For many of them society requires either that 'red blooded males' conform to an increasingly restrictive behaviour set, or that less red blooded types are conform to an idealised and 'symbolic' kind of manhood that complements that of the symbolic female.

---

[4] ONS (Office of National Statistics)
[5] Bullying. European Psychiatry Vol28 issue 8 pp498-503.

It begins early. At school, these stereotypes heavily influence behaviour, acting as markers in exploring roles, relationships and responses. These early trials of personality are rather arbitrary, as it coalesces around the abilities, character and early sexual psyche of the individual. Mistakes happen, causing embarrassment or even wholesale and indelible grief. Males are supposed to be robust, but evidence shows that both males and females, in this exploratory pattern of development, experience lasting hurt. Any kind of bullying can have lasting consequences[6]. In some cases, bullying at school correlates with increased incidence of being unemployed (or socially side-lined) and, thereby, with the risk of suicide.

The world we have constructed, albeit unknowingly, creates severe tensions and anxieties, engenders depression. No other animal commits suicide.

Now in early adulthood, our young man has to navigate a critical pathway; society requires him to conform to socio-economic expectations. Many do this as their 'day job'. When 'off the leash' they might reveal a different aspect. Some have a quiet passion for video games that show hidden talent; fast reflexes, high levels of concentration, aggressive and focussed behaviour, ability both to handle risk and at the same time wholly embrace, even enjoy, its inherent dangers.

If this sounds like the profile of a manager of a hedge fund, the likelihood is that the one actually holding down

6

the job is tired out and resting quietly at home. We might imagine his primitive ancestor returning from a stressful and dangerous hunt to a quiet evening with the family group and falling asleep at the fireside.

Meanwhile our gamester male might be seeking challenges that are absent at work; the biggest threat is the guilty feeling induced by utter boredom and the threat of losing his job. Endless repetitive routine is not attractive to men. In the past, many would prefer the frisson of a call to strike. Anything to break the monotony. Thus, robots are better suited to repetitive tasks, and much human talent wasted in dividing work into constituent elements in the name of 'efficiency'.

Young men are 'elsewhere' a lot of the time. A few million years of conditioning to high levels of risk and life 'at the limit' is sufficient to explain their frustration; their physical energy is wasted and their evolved body shape redundant in a modern context. 'Society' tacitly accepts the benefits of modern society because patriarchal organisers profit most. Individuals chafe against the encompassing yoke, searching for compromises. The 'work-life' balance acknowledges the conflict; that 'work' is definitely not 'life'. For the hostage male, for a large part of the time, most of life is 'elsewhere'.

There has to be early conditioning within a stable family to lay the groundwork necessary to funnel testosterone and adrenaline-fuelled behaviour into channels of compromise. Sport is an acceptable and healthy alternative to joining a gang or selling drugs. Many team sports mirror exactly the probable behaviour of the hunting band.

Meanwhile, the controlling patriarchy explicitly encourages risk-taking if it benefits society or the economy, but you have to be at war or to have lax financial regulations to exploit male sexual psyche to the full. Money diverted to sport and entertainment has the dual benefit of providing a healthy return and channelling male testosterone.

An evolving niche has adroitly managed to exploit both ancient skills and the linear logic of the hunt to create 'pseudo organisms' in the shape of corporate entities. We can safely categorise these organisms as having pseudo existence; while directors, managers and staff come and go, die or retire, as investors buy and sell the shares, as premises are sold and new ones acquired and the seat of business migrates, the business or government agency plods on, sometimes through centuries.

Companies, corporations, institutions and organisations are proud of their lengthy heritage. They operate as 'clans'; transient members for come and go, but the name or identity survives, outliving the individual humans who benefit from, and participate in, their success. Cadbury, Schweppes, Lloyds, Warner Brothers, Penguin, Balliol, Harland and Wolf, Ford and many more behave in some ways as living entities; bent on survival, acquiring the raw materials necessary for their survival, organising protection for 'family' members from the ravages of the 'world outside'.

Hierarchical families and established criminal gangs operate in the same way.

How is such behaviour 'hard-wired' into modern humans? Of course, a stalwart male (of any species), driven by hormonal urges, seeks out a female. The same drives stimulate other males. Competition is mostly 'for show'; it's not worth getting hurt and it's better to back down if you're not 'top dog'. Choosy females select Dominant males. In modern society, it's very simple, this means the one with the most expensive car; important symbols confer status and therefore imply social dominance (even though the driver may be an ugly little runt).

Breeding, in ideal conditions, weeds out the ones who have the poorest chance of breeding, so types with dominant

Biological, chemical, experiential, environmental factors may influence neuronal 'fate'; as the brain develops, mythylation ('chemical change') can result in either activation or repression of gene transcription. Because these are essentially Neural Stem Cells, their final destination or function is not entirely 'fixed'. Position and final form of Neural Stem Cells may impart behavioural traits according to these outside (epigenetic) influences. Three stages of gestation - early, middle and late – may be crucially different as to the 'fate' of NSCs. So, the life experience of the mother during these phases has an effect on later behavioural and health outcomes for the infant, influences which may persist throughout life and are difficult to modify. Researchers point out that persistent environmental and behavioural patterns in parents influence the health patterns and behaviour in children. Pockets of social deprivation tend to persist if neuronal development in children is heavily influenced in this way.

See, for example: Behavioural Epigenetics, Greg Miller. 'Science' 2010 July.vol.329. No.5987. pp 24-27.

characteristics are born to choosy mothers. Recent findings suggest other factors come into play during gestation; offspring may acquire behavioural traits in this period, other than physically inherited characteristics determined by DNA chromosomal sequences (see panel).

We can investigate how this might happen in a male with a primitive type of sexual psyche. We've seen how the male in this sense is hostage to his nature. It can be difficult to modify. Indeed, patterns formed during gestation, infancy and childhood can have such disastrous results that it is astonishing that the world is mostly full of decent, affectionate, caring, responsible males.

But, let's take an extreme example, the worst case scenario ; Adolph Hitler. Male, powerful, attractive (yes, women adored him), an archetype of patriarchy. Many thought he represented the essence of male dominance, combining leadership (the 'Fūhrer') with a fatherly concern and a neat line in propaganda and poster art.

His message was 'identity' stamped (sometimes literally) on the consciousness of an entire nation. In the tradition of ancient Rome, in the footsteps of Genghis Kahn, in Europe, Hitler ruled supreme.

He had many of the primitive characteristics outlined previously – harnessing the principle of action at a distance, projectile weaponry, membership of an organised warrior group (in this case 'storm troopers',

210

Hitler's version of hunter-killers), a holy clan, armed and dangerous. The women loved it.

Hitler was their icon; lauded, lionised, idolized; an ideal father figure and protective husband. He emphasized Germany as the 'Fatherland' or the 'Motherland'. Heinrich Hoffman, his own personal photographer, ensured he was pictured alongside attractive children. He also imparted to his regime a powerful male ideal by organising endless rallies featuring hordes of marching men. It's happening again to-day in North Korea, and the women are joining in.

Key to his powerful image – a completely fraudulent one – was organisation and 'set dressing', even down to goose-stepping; the exaggerated straightening of the leg, imperious gaze, the rigid discipline was thrilling at the time, though not to everyone. At their head was the personification of the hunter-killer in a modern industrial context.

To indicate just how flawed is this view of the male personality in its most extreme (indeed, insane) incarnation, a review of his childhood reveals a great deal.

How far had he come from being a 'normal' male? Despite his power, Hitler was very far from being normal. The 'ideal-ego reaction formation' was finally diagnosed by the OSS, the US Office of Strategic Studies, the forerunner of the CIA. Recruiting a Harvard academic, the OSS drew up a classified report of Hitler's history and the probable cause of his desperate drive to dominate; his aggression and thirst for revenge.

Worst, was the compelling need to project 'criticisable elements of self'. That is; he *felt* inadequacies and failings of childhood so strongly that their insistent presence he necessarily projected onto others; he loathed people. His father was brutish and treated his mother brutally. He despised his mother's enforced weakness. Fearing and admiring the power and brutality of his father, hating his mother's unending humiliation, young Hitler grew up hiding a welter of compulsions woven with 'stubbornly suppressed feminine sentimentalities'.

Not present in the diagnosis is another symptom not documented before; Hitler's hatred of the Jews emanated from his association with repressed ideas and the fear of women. I suggest a primitive aspect of male psyche consists of an innate fear as previously discussed: women represent a mysterious 'natural force', one associated with 'mother nature' and unexplained 'natural forces'. Men worshiped these forces, variously, as deities, or feared them as being all-powerful and beyond their control.

Hitler's appearance here is to present an extreme example, *the* extreme example; the deceit, the cruelty, the mad energy in the attempt to annihilate an entire racial group, this, because Jewish people represented an intelligent, gifted, creative, sensitive, diverse and unfathomable cast. The OSS report failed to consider this aspect of Hitler, but an awful truth lurks in the case history.

Hitler's mother was 23 years younger than the brutal, tyrannical father was. His first real attachment in later years was to the daughter of his housekeeper, a young girl when she first made an impression on him. A relationship of some kind occurred. We only know that his housekeeper's daughter was nineteen years younger than Hitler. She committed suicide. Eva Braun, his last 'companion' committed suicide also, alongside Hitler at the end. She was twenty-tree years his junior. Another 'companion', Reiter by name, was a young lady twenty-one years his junior. In 1928 she attempted suicide but failed.

Hitler, contrary to the fantasies of contemporary females (and squadrons of patriotic soldiers), was an infantile, compulsively extreme, sexually impotent individual, suffering moments of 'extreme self confidence' alternating with 'homicidal compulsion and brutality'.

Hitler was a man who began life as a frail and sickly child, emotionally dependent on his mother. He failed to demonstrate, ever, male potency in the presence of a woman. Yet key drivers in the expression of his 'homicidal compulsion' was love of his 'motherland' and that Germany would become the ultimate 'fatherland'.

We can see how a derailed male can still elicit love, admiration, sexual responses, entirely the result of outward image and display. Those who knew Hitler, such as Albert Speer, remembered his cold words, his "preference for unintelligent women, who could not challenge him" or "disturb him when he wanted to relax".

He is here, briefly resurrected, to demonstrate, not the base character of men, but the ominous character of patriarchies; here was a man riddled with neuroses and certifiably insane, yet he became the lead individual of a patriarchy that dominated Europe and was responsible for untold misery, slaughter, persecution, torture and mass murder. Patriarchy is, and has throughout history, been the scourge of humanity. Patriarchy is the result of humans not realising or accommodating the personality and sexual psyche of women, to which they react in fear.

Insanely directed patriarchy is a recurrent theme in world history and persists today. Many male tyrants enlist the aids of like-minded male comrades that sanction his behaviour. Supposed heroes (Henry VIII, Chairman Mau, Bill Clinton) turn out to be less, in their ordinary persona, than their public had   expected them to be. It is the more primitive of male characteristics that often come to dominate a patriarchy, not the noble, generous, just and honest qualities of our aspirations.

Many societies decay and collapse specifically because they are patriarchal. The kind of male leaders with sufficient internal motive to aspire to lead are by that very definition, unfit; they are male and the ambition is visceral in origin. Perhaps the more cerebral types have no stomach for it.

How lethal patriarchal organisations can become is well-documented. Patriarchies simply happen to take over large-scale human systems because they are most 'fit' to do so; collectively or individually, males have the credentials, honed through millennia. However, that they are the recipe for modern states needs to be questioned. Even in democracies, complexities of our present era seem to demonstrate that a patriarchal hierarchy, with attendant risks, is unlikely to deliver stability and well-being.

Are we to allow 'natural selection' to deliver large-scale organisation of human society when it now becomes obvious that humans are an artificial adjunct to a natural environment? We use our collective morality, whatever its origins in the misty regions of human history, to formulate a 'moral' environment, not a 'natural' one. Heaven forfend we should behave as our natural instincts demands; of all the Earth's creature, humans are the most beastly.

That we are captive to our male and females roles is indisputable. Even at the blurred edges, few roles are convincingly 'neutral'. But, as we shall see, that may be the direction in which what I call 'Neoteric Evolution' may propel us. Being hostage to fortune – if we can survive that long – may even turn out to be fortunate.

CHAPTER 14

Nature as the Enemy

All males reflect their primitive origins. It was, after all, only a short time ago that tools, especially stone tools, turned us from feeble foragers into lethal killers. Perhaps this fixed in male mentality the notion that all nature is a threat. The modern ideal, since it was first articulated, has been to 'conquer nature'. See also; *'conquer space', 'conquer the sea', 'conquer the mountain', 'conquer the wild', 'conquer the desert', 'conquer the forest', 'conquer the deep'*, etc. The Arctic, the heights, the depths, all of nature is the enemy; we use the language of war, struggle and conflict and unwittingly reveal our hidden psyche, or rather, not so hidden.

Much of Hitler's temperament was shaped by the need to project those drives onto his world and, ultimately to destroy it. Clearly, few men are the Hitlerian type. His appearance is only temporary, again to emphasize the extreme and rare case. But the analogy, as it now becomes, serves to demonstrate where the primitive fears and their deleterious projections still form the well-spring of humanity's flawed relation with that other niche, the 'natural world'.

We have built our own niche; recall imbued us with the capability to make internal decisions about how to shape the world to suit us, not the other way around.

As we can see, Darwin and subsequent science is blissfully unaware of this fact. So, too is governance and economics. In fact, the latter prefers to ignore the dilemma this presents; business often shows the same Hitlerian projection; nature is mysterious and unfathomable, let's ravish her.

The justifications are, variously; human well-being, progress, the conquest of 'poverty', 'fighting' disease, exploitation of natural resources for the common human good, peace, profit, defence, 'conquering' hunger, defeating inequality and other chimeras.

Poverty, of course, is relative; is an aboriginal group deep in the rainforest of the Amazon conscious of being well below the 'poverty line' or 'suffer from inequality'. Should they, as many westerners believe, be rescued from their 'condition'? The same goes for disease and inequality. These are modern chimeras. A few thousand years ago (recent history in real terms), all these modern preoccupations had no meaning. Do our primate cousins suffer from inequality? Our humanitarian, political, emotional, intellectual, social and sexual preoccupations don't bear scrutiny. They result in the very conditions we have ourselves created; over-exploitation, excessive expansion of our niche and dominance over others species in the biosphere.

It goes without saying that patriarchal structures directly reflect a distinct pattern of male attitudes and behaviour. Most organisations, as well as nations, reflect this in their structure, motive, indeed 'raison d'être'. They are aggressive in defence, acquisitive, exploitative (of human and natural resources), expansionist. In the case of a commercial organisation the comparison with other species is most apparent, which is why (mentioned earlier) we can regard them as pseudo species.

We can now add to the pseudo species the very real characteristic of patriarchy; an organisation dedicated to everything characterising the male in its untrammelled form: linear logic, following a set plan to attain a series of objectives, drives it forward, as effectively and as speedily as possible (that is, economically or with least cost to itself). For commercial purposes, ruthlessness is a key characteristic in a member of this type of pseudo species. As with our hunter-killer band, there are others 'out there', some of them dangerous, after the same prey; this is 'the competition'.

Let's now add in the Hitlerian, 'off the scale' measure of male-attributes, combined with an economic identity, that's 'out of control' and we have the very conditions seen in collapsing economies and banking systems in the 19th, 20th and 21st centuries. Risk-taking and aggression characterised these periods, sometimes to the point of self-destruction. In his report compiled for the OSS Dr.

Henry Murray predicted Hitler's suicide long before Germany's defeat.

One can imagine why aggressive, risk taking males are prime target for recruitment in the financial markets and why financial controls are now a formidable (we hope) deterrent to the instinctive behaviour of this new breed of financial hunter-killer.

No wonder women are attracted to dominant regimes awash with wealth and male testosterone. The heady atmosphere of power is a strong influence to join with them as a substitute for the more basic drives of the sexual psyche. It may be that this part of a woman's personality has bowed to the increasing pressure of patriarchal influence and her psyche modified. Maternal instincts are certainly either actively or aggressively discouraged.

The identity of the female and the values she holds are increasingly at odds with the work environment. She can't reasonably be expected to become a surrogate male; her psyche is both more 'naturally evolved', by which I mean more ancient, formed long before recall produced tools capable of taking on prey not normally on the humans' 'shopping list'. Little or no satisfaction devolves to them; their sexual psyche becomes redundant and subjugate in these environments. Their sensory-intuitive nature, breadth of intellectuality rather than the narrow linear logic of the male, demands stimulation and rewards of a different kind.

The evolution of humankind has progressed to a point where pseudo species are now the dominant form.

In this context we can pose questions that highlight a human, not just a female, dilemma, though hers is more acute. Lacking a voice, (in each constituency, a female parliamentary candidate for every male is not beyond reason) female ideals and creativity need a forum not dominated and encircled by the patriarchy. Where is female culture? Where is her music, her art, her literature, her theatre, a spread of TV channels other than the patriarchal–defined frivolous, sexy, fashion, cooking and product orientated? Women are not just consumers.

On the important physiological context underpinning her consciousness, society is silent. Patriarchal society excludes childbirth as a celebration deserving more than a greetings card. Were it a male phenomena, we'd celebrate it as a heroic challenge, offering bronze medals, mentions in despatches. Women (like Lady Fanshaw) make light work of the emotional storms and physical discomfort, of downright agony, heart-rending loss, of menstruation, pregnancy and childbirth.

Instead, her role is characterised as something needing immediate medical attention. Apparently, in the male-dominated world, being a woman is akin to having a life-long illness and through emphasising science, men skirt around the problem.

It was not always thus. Yes, pregnancy is a physical trial involving sickness discomfort, pain to the point of

Recall

This is indeed a powerful weapon. The recall paradigm drives humankind through the powerful, aggressive, resourceful male, further evolving the species. In the natural environment, the logic that drives him is absent. There is no logic in nature, unless we admit to a deity with a human outlook. Instead there is only the information we have accumulated. We started with Sumerian trade goods and in Mesopotamia. Professor Schmandt-Besserat of the University of Texas describes maritime trade and its method of accounting; on a clay envelope, an orange shape indicated oranges in the cargo. Inside the 'envelope' itself were the required number of little clay oranges. Later the clay tokens were dispensed with; the envelope became a tablet. Writing was born. With it came 'information'. Science errs because of 'false accounting'. We make the assumption that each orange is alike because we are deluded by recorded data, assuming it reflects 'reality'. Whereas, oranges snowflakes and everything in nature, down to sub-atomic particles, are never constant, never 'equal, never, ever, 'the same'. Recall creates the illusion.

agony. Many of our ancestral species cope well. Nature copes and humans have lost that ability. Pregnancy and childbirth must be celebrated, and positively championed. Football, cricket, motor racing all have a higher cultural recognition than childbirth; no laurel wreaths for champion mums or prizewinning midwives. It is why women need their own political candidates (women), their own parliament, with representation equal to that of men in law, spirit, effectiveness and stature. Surely, they couldn't do a worse job? The patriarchal system leads to a gross imbalance, failing to reflect the true nature of humanity. The male sexual psyche is the greatest impediment to these ambitions. His linear logic, a disinclination to weigh alternatives or deviate from a set of actions bedevil him; a male has difficulty in revisiting ideas to look for flaws or alternatives. Men are notorious

221

for 'digging their heels in'. 'Carry on regardless' is the favoured method despite mounting evidence he might be in error. The inevitable consequences evoke irrational anger. As we may conclude, man is at odds with nature. His default position is to do battle. He finds no logic in nature and thus, as Murray would argue, must project the logic in his own head onto essentially chaotic processes. Thus, triumphantly, man declares to have discovered order. What he discovers is that recall has induced order in his imagination, the power to imagine the future based on past circumstances.

Finding no order in nature and he strives to conjure it and in doing so masters science, or rather, science becomes *his* master. For our female, trapped in surrogacy to the male patriarchy, conflict becomes part of her life, even the norm.

She can deal with nature, empathise with its disorder. Each monthly cycle is a new experience reminding her of the power and uncertainty of nature throughout her reproductive years. She is part of a 'primitive' world, just as every other species, every living creature, is part of a 'primitive' world, one that requires no explanation, but which ineffably sustains us.

Male expectations imagine nature as something to conquer, as we have seen.

Man is deeply uncomfortable with the notion of femininity because, though an essentially natural component of human life, nature itself is alien, an enemy.

Male instinct, confronted with any phenomena and its chaotic character, is to 'crack it'. But, throughout the history of scientific discovery, the solution proves to be untenable and is always superseded.

Remember Nietzsche; " Logic rests on assumptions that do not correspond to anything in the real world."
He could have said that nothing in the real world corresponds with the assumptions of logic. Logic does not 'exist' in nature but is fabricated and adjusted according to what is self-consistent. The course of history is littered with abandoned proofs. Logical explanations turn out to be false, or the logic subsequently adjusted to accommodate a proof. Prior logic is abandoned if it turns out to be no longer self-consistent, even though once it was irrefutable until the emergence of new evidence. Science and human enquiry is unquestionably a 'moveable feast'. Beyond the catalogue of 'established fact' lie further mysteries, as every almost (invariably male) Nobel prize-winner knows.

CHAPTER 15

The Course of True Love...

After a tiring and haphazard passage through early life, our two young adults face challenges that their forebears never encountered. Technology and social change represent an obfuscation of choices; decisions of an earlier century were simple. Today innumerable complexities conspire to hamper choice; lifestyle, opinion, fashion, cost, choice, status, risk, credit rating, friends, time, location, travel, education, work, career, housing, mobility, health, appearance, sexual orientation, taste, acceptability, leisure, gender, age, diet, race and a good few more.

Earlier epochs demonstrate little evidence of choice; you were what you were and who you were: in the slow pace of change, instilled was acceptance. Religious belief forbade questioning of the hierarchical order, a patriarchy with a male god 'at the top'.

However, whereas choice is now bewildering, the skills for making choices have been eroded. This used to be the work of parents, but a patriarchal system undermines parents, who are not to be trusted, it seems, with the supply of compliant citizens. Instead, education falls increasingly within the remit of the state. Yet, in the 21st century, schools are failing; boys, especially white boys, fall significantly behind black male youths and all girls in education in the UK.

A teaching workforce of whom 80% are female dominates state education. This breaks down to 85% of primary teachers being female and, in secondary education, 62%. Males suffer a disconnect and it is odd that white males do worse than black; the proportion of black youths from low-income families gaining 5 or more GCSE's in A*-C grades (2015) was 43% and for white youths from the same background just 28%.

This bias (Equality and Human Rights Commission, 2015) comes courtesy of state institutions supposedly committed to equality in education and all other spheres of influence by the state. But states have other agendas unspoken of or unrecognised. Females – 'prone', in the language of misogyny 'to pregnancy' are being weaned from their 'natural' role in a manner that seems to confirm a change within the species; a trend towards diminishing sexual reproduction as the driving force of natural selection in the human species.

In this new pattern our young lady (sorry, 'woman') will do better educationally; young women are now more likely to enter university than young men. There's a changing gender bias in degree courses, too; in 2007 34,035 more women than men were at university. In 2015 this had risen to 66,840, an increase of 96% over an eight year period. Women graduates, now outnumbering men (56.2% women against 43.8% men) from thence enter a more diversified labour market.

But all the signs are that, while society, business and the state all benefit from a more diverse pool of new entrants, the costs to the individual male or female are becoming hard to bear. Raising a family is a distant prospect and any child or children born will be born to much older parents than earlier generations. True, they will be 'valued' more, 'cherished' even beyond measure. The 'value' now of an infant is immeasurable in comparison to the time when Elizabeth Sanger's mother gave birth to eighteen children in twenty years.

Of course, children who survived the uncertain course of infancy are highly valued. Once, the affection and care was tempered by the near-certain knowledge that survival of an infant was never a certainty. Today, a family with 1.9 children lavishes attention, smothering a child with affection and an excess of stimulation.

Yet the modern environment has hidden dangers to equal that of infant mortality. Unwary members of our societies face depression and other forms of mental illness practically unknown in less developed worlds. Stress and complexity of modern states are a continuing threat to the primitive male and female sexual psyches. Our basic abilities are grounded in physically evolved forms. Our abilities exploited by the modern environment are extensions of mental evolution, based on recall.

There are no apparent limits to the expansion of the mental niche. Our intelligence is now both artificial and based in machine technology. Our young woman graduate will find work in which computer skills are essential. It is taken for granted she will have already reached a standard of computer literacy in parallel to her extensive and lengthy education. She might be employed in programming, adding to the exponential growth of the mental niche.

There are other costs besides delayed and perhaps artificial parenting. Illness unknown to primitive cultures afflict our modern generations; the bare facts are that women are 29% more likely than men to be treated for mental illness, though this bias may be due to male under-reporting or failure in diagnosis.

Depression is also more common in women, affecting 1 in 4, while depression affects 1 in 10 men.[7]. By 2020 depression will be the 2nd cause of global disability. In men, dependence on alcohol will affect 1 in 5.[8]

The rapid and extensive advance of civilisation exacts a price. On individuals, males and females, there is a heavy toll. In a sense, the mental niche requires increasing cerebral and less practical skills. Women increasingly gravitate to work involving the skills they innately reflect. Males, on the other hand, have less opportunity to exercise their innate skills. Primary and further education does little to encourage a sexual psyche that is becoming increasingly redundant. There is little need for the mechanical and industrial skills at which they excel. As male personality becomes overshadowed,

---

[7] National Institute for Clinical Excellence.2003.
[8] World Health Organisation

their identity is increasingly challenged, their innate abilities less valued and less valuable. The increasingly cerebral world of Intelligent Technology expands but needs only small numbers of specialists and application designers.

In one sense the sexual psyche of the female is being 'masculin-ised', to suit both patriarchal organisation and a falling demand for new humans. Industrial activity is declining and the need for mass muscle-power, working in well co-ordinated industrial settings, seems antediluvian. No wonder, with such slender support for male identity, suicide is now the prime cause of death in males at 35 years of age, surpassing females – in spite of the latter's widespread and well-documented depression – by nearly four times the female rate.

In truth, physical evolution cannot match the speed of expansion of the pseudo niche. Each computational or technological advance instantly 'feeds back' to the mental, imaginative, creative energy expended. Exponential development is leaving mere humans behind. That 'social pressures' also feed back into employment, educational and community structures is unsurprising. The slowest to change are those caught in the early 20th century time-warp of heavy industry, heavy drinking, poor housing, poor health and poor education.

We can project these effects onto future generations and imagine how violent the upheaval will be. Advanced ('quantum') computing, the internet and information technology already by-pass swathes of a population who simply do not 'fit' the new model of social organisation.

The social and commercial pressure - as our young woman in her first commercial role is finding out – is to remain single, spend income on the goods the new economy offers, enjoy a lifestyle of entertainment, good food, avoid carcinogens. To see exercise as leisure (there being no natural demands on the body to burn energy) and sift the preoccupied male population for a possible candidate to engineer an acceptable mating outcome.

Young males may have to revert to primitive behaviour of the hunter killer in search of commercial success, or tone down their behaviour, become more feminine and receptive, learn sensory intuition or combine these facets within a chameleon-like character, perhaps enjoy sexual liberty until premature middle age sets in at 35. Oh, and avoid alcohol, drugs, fast cars, and extreme sports. Or not.

These two profiles are outside of what might be called the specialist caste. This caste will be born to already well-schooled parents, who have avoided cosseting their offspring. From these beginnings, an independent mind plus challenging education are the basis of a career that adds to the structure of the mental niche. Each further advance accumulates greater value, so the rewards to the man or woman (perhaps the planned offspring of the two examples now constituting further modern pairings) are rich indeed and new financing and insurance patterns will ensure accumulated wealth, property and an early retirement.

For others, taxation will provide for their basic needs but a widening gap will follow the pattern of the early 21st century. The 'Haves' will have everything they could wish, for there will be fewer of them. The gap between this caste and the 'almost have' will be large, the financial gap measured in millions, if not billions. The 'never had's will form the third, resentful and unruly class who will refuse to comply with the usual social injunctions to 'behave, whatever the circumstances'. They will form the largest group, consuming the 'virtual' goods engineered and marketed by the two higher castes.

The 'never-had's will be the product of a never-had-it-so-good culture in which computing and robotics have displaced them. Leisure and entertainment, eating and drinking, and general enjoyment will be all that is left. The fundamentals of consumerism will be called into question. The question will be; 'there must be more to life than this'? They've never had a life.

What identity will look like, how the two different sexual psyches will measure up is hard to predict. How far two psyches can be sublimated to, or subsumed by, a cultural change, is a subject for further conjecture.

CHAPTER 16

Eleven Miles High

The expanding mental niche relentlessly picks away at human identity and any remaining differences between males and females. The sexual psyches of humans – as we have seen – have a respectable track-record; they served us well throughout more than five million years of prehistory. And sexuality is not just the mammalian model for successful reproduction. Many species evolved in the same manner, with sexual characteristics linked to selective processes; colour, size, behaviour; all contributing to the (relatively) smooth process of reproduction and survival. From apparent chaos, patterns emerge; types become distinguishable, species differ and members easily differentiate.

Humans began to evolve differently; our mental architecture diverged from the animal line perhaps 5-8million years ago. The burden of 'mind' enables further specialisation. There are commensurate difficulties, which, in turn, need a quick mind. Humans quit the Darwinian model of evolution long ago. Ever since we respond not simply to natural cues from the outside, we act also upon cues from the inside. Imbued with this facility of reflection and anticipation, all kinds of alternative choices present themselves. For the female, the prospect of an apparently attractive male - burly, dominant, respected by other males and top of the local hierarchy - at first seems an attractive choice, appealing to 'basic instincts'.

When duly considered (that is, recall gives an alternative and, in effect, widens the choice), a thin, wiry, dull-looking male who has shown in her memory and experience (recall again) to use his brain rather than his brawn to conjure long- lasting rather than short-lived outcomes, will turn out to be a better choice.

We can see how the language presents the recall paradigm as the prime -mover of human behaviour; 'duly considered', 'to conjure', 'long-lasting', 'short-lived', 'outcomes', 'will turn out'. No animal of any other species has this resource; all species respond to a direct stimulus, they can't help it. From minute to minute, animals exist. We can't help what we are.

In the human species, a combination of conscious behaviour and 'animal' responses result in a unique sexual psyche. The human male and female, though crucially distinct, unlike other species, engage consciousness to delay responses, consider alternatives, plan outcomes more favourable than mere response.

True, this new regime can be bypassed in the 'heat of the moment'. Instinctive attraction, 'passion', is enjoyable because it frees us, however briefly, from the burden of mentality and reason. Reflection no doubt 'kicks in' the 'morning after'. Reason adopts its inevitably plodding and insistent course.

This hybrid creature – mankind – reacts to stimuli but also has a unique ability to 'step outside' the chain of endless reactions that characterise the rest of nature, which is a matter of thoughtless responses to random events.

It is humans alone who demonstrate evolution may itself be evolving - a kind of hybrid mix of conscious and unconscious reactions, set in an artificial niche; the ever expanding and evolving product of 'mind'.

From 1991 to 2009 IVF treatments increase by 700%, averaging a 38% increase each year over an 18-year period.

(Human Infertility and Embryology Authority)

Thus, the sexual psyche is in flux and humans may be in the process of evolving into a 'sexless' species. We can see that humans are already following a path of mechanised and 'medicalised' sexual reproduction. We may cling to an ideal of natural or sentimental pairing, but in modern encounters we insist on safeguards and oversight.

In liberal sexual parings, protected sex eliminates dangers and 'risks' of pregnancy: not long ago 'risk' in pregnancy implied death in childbirth for infant or mother, or both. Elizabeth Sanger's mother was one of

233

the lucky ones; times change. Times change so quickly that soon couples, or an individual, may be able to choose, select (or buy), a male, female or sexually neutral individual, to nurture and educate them, with no need to get involved in any messy sexual context. One day, the very idea might seem abhorrent.

Further advances in human evolution will be strongly influenced by concerns over population – not just the raw numbers but, at the top of the patriarchy, by quality, type, skin colour, sex, intelligence, talent, height, eye-colour, temperament. Once the human genome can to be re-assembled at will (supposedly by well-meaning and rigidly controlled laboratories), who can say enterprising commercial groups will not emerge, providing a choice of real 'designer babies'?

Moral justification for science to follow this course will no longer be any more meaningful than that of commerce; 'merely satisfying consumer choice'. It can be convincingly demonstrated that science is, after all, as it was admirably written "In truth, this was the century in which science became the new form of religion" and science is always presented as " matters of unassailable dogma". This is as true today as it was in was in the 17th century[9] Nothing has changed since then, even though it is equally easy to demonstrate that science is, like religion, merely the worst of many figments of the human imagination. The history of science is a catalogue of advances, once the arcane preserve of the investigator, now translated with indecent haste into commerce. Let's

[9] Ackroyd, Peter: 'Rebellion'. 2014

hope, in the case of chemical or mechanical human reproductive processes, there are not too many mistakes along the way.

Meanwhile, human sexual reproduction may become increasingly constrained, the last vestige of chaotic nature impinging on human behaviour and perhaps rather naughty. Our future young woman and young man may enjoy uninhibited sex with their respective partners yet choose designer offspring – male, female or asexual/neuter from an online-catalogue, re-assembled from selected sequences of DNA, according to taste, a bit like a takeaway, only human, and with the prospect of being impossibly bright.

As this advanced culture accelerates away (at an increasing rate) from its natural origins, each member will be highly prized and have huge inherent value (given the duration and complexity of its development). This high value will have to be protected; each new individual captive to a predestined role – a "Pharaoh" culture in which each individual is going to need extensive and constant surveillance (IT) a great deal of (mechanized) support.

Even more precious will be a fast disappearing 'natural world' and nature a threatened reservoir of valuable bio-chemical treasures. Not all the castes of humans will be asexual or neuter. There may be 'revisionists'. After all, a huge reservoir of documents and imagery will be witness to a time that may be strongly attractive to those from whom instinctive behaviour has not been entirely eradicated; perhaps women will be in the majority.

The circumference of the Earth is 25,000 miles.(or roughly24,901, and it varies slightly). The film of breathable air atmosphere above (the tropopause)is just 11 miles thick.11 miles represents 0.0441 % of this circumference, a miniscule fraction. Think of an orange and its paper tissue wrapping – it's the same kind of ratio. Given that most of the Earth's surface is sea and much of the land surface mountain or desert, our biosphere is perilously fragile and severely limited.

Reverting to a portion of the natural world reserved for leisure or science may one day constitute a privileged holiday; tomorrow, you can book a cruise, staffed by specialists in various sciences, to visit far-flung and nearly inaccessible corners of the world where nature still manages to flourish. They are disappearing fast. The privileged can watch vast icebergs calve as melting of the ice caps progresses apace, or witness, like sad voyeurs, the dismal sight of mountain gorillas in the last remnants of their once pristine territory. We should hurry.

Natural calamity is a factor in Earth's history with a remarkably constant presence. Recent human history – five or ten thousand years, even a million or so – is a mere dot in the sweep of time encompassed by evolution. Microbes and bacteria have survived for four billion years. But, like the dinosaurs, it's the currently 'advanced' creatures that perish. Whereas we are still relatively primitive, our projected systems are not. We

236

rely heavily on science and tech-based systems that are vulnerable to disaster. A good flood can wipe out our most valuable infrastructure and we will be powerless - literally.

So there's a good argument for us – or our young man and woman of a future age – to skip directly to the revisionists stage while we can. We have to put these outlandish fancies about the future in context - and here is where it gets serious.

It is a common misconception that climate change – one of those calamities that might or might not befall us – is a distant, long-term or remote problem. The reverse is the case; the atmosphere of the Earth, the breathable zone (the Troposphere) is only eleven miles thick, and only about the lower two-thirds is a useful climatic zone of breathable air. It's advisable to imagine what that thin vertical strip looks like laid horizontally; a there and back trip to visit friends, to go to the town centre perhaps. Eleven miles is halfway from central London to the suburbs, the kind of commute many people 'living in London' do twice a day. You could walk it. Many cyclists cover the distance in under an hour.

You can see France from the cliffs at Dover. Half way across is equivalent to where our biosphere ends.

Put another way; imagine an orange covered in thin tissue wrapping. That's about how thin is the layer of the Earth's breathable atmosphere. Other sobering truths in an equation aimed at 'guesstimating' the near future of humankind must include difficult sets of numbers; firstly population growth: in 1800 the Earth's population of humans reached an estimated one billion. We had taken perhaps five million years to reach that figure. In the next 127 years, in 1927, we had added another billion. Recently, in just 14 years, the latest billion has been added. It is now around 7.4 billion, and counting. Projected figures are 8 billion in 2024 and 9 billion in 2048. My, how time slips by!

The next set of figures concerns useable land area; 71% of the Earth's surface is water. Only 30% is dry land – sometimes very dry. Of the 57,308,738 square miles of land surface, 33% is desert and 24% mountainous, making 57% uninhabitable, leaving 24,642,738 square miles to accommodate 7.4 billion people (about a third of a square mile each).

Portioning bits of land to individuals is meaningless, however. The trend is towards intense urbanisation; in the UK today 87% of people live in urban areas, in concentrations bordering on hazardous. If sea levels rise, gently sloping coastal lands, the site of much agriculture and many urban conurbations, will be the first zones to become uninhabitable.

Throughout history, societies in competing, patriarchal groups initiate war and conflict. Outpourings of primitive masculine responses left millions dead or displaced. Of those affected by violent conflict, civil wars, natural disasters and displacement, amounting to 50 million people, 80% are women and children.

From Stalin to Hitler, from Louis XIV to George 1V, from Hitler to George Bush and Donald Trump male stereotypes dominate the landscape. Endless lists of male names describe the heroic and the fiendish, peacemakers and the insane, all male.

Women don't have, are not allowed, cannot be ascribed even a family name that will last through generations. They, meanwhile are the bearers of our children, the future populations. Over time, these children will come to be a valued resource, more valued than throughout human history. Yet in concert with the importance of an increasingly threatened humanity, women become relegated to being mere onlookers, passive participants in a male-dominated social, scientific and technological world heading for self-inflicted global disaster.

The challenge will be to find a means for humans to survive 'man-made' catastrophe. Science, as always, will fail; there are no solutions in the restricted, illusory world of linear logic. Nietzsche was right; hollow outcomes of particle physics, medicine, earth sciences and IT, all demonstrate an inability to 'order' survival or peace, engineer sustainable economies or environmental balance. They strive toward 'new' methods, not revive old ones that we know actually work.

There are around 10-14 million species occupying the fragile and finite biosphere described here. One alone threatens them all. Of the 'all', we haven't even been able to fully investigate an estimated 80% of existing species. Of those identified, a large proportion has yet to be described.

What characterises the human is arrogance, especially the human male. Recall imbues consciousness with an illusion; that time flows through nature. This illusion allows the human to modify the niche to better suit its occupant. With all other species, without exception, the animal instead is modified, as Darwin described, 'by slow degrees', better to suit the niche.

Thus the human is, in a sense an interloper. We can only account for this by admitting that, with the human, evolution itself evolved; that 'natural selection' is no longer its principle driving force. Much of the natural environment is at the mercy of human interference, is at '*his* disposal'. Mankind (this time I purposely use the male prefix) has become the ultimate patriarch. All his previous attempts to enshrine various 'gods' as ultimate authority were merely a psychological reticence; science and recall now legitimise 'maleness'. In the modern age, he now considers himself, quite literally, Master of the Universe.

A greater danger can now be clearly seen; a burgeoning population of humans with no natural enemies except microbes (and ourselves). There are no limiting influences since recall suggested tools and we shaped out own mental niche. Science replaces religion as the mind of man now presides over motive, behaviour and destiny.

The mind of man is in thrall to a still- primitive set of drives, spurred on by the success of linear logic in describing and taming the 'forces of nature'. Ambition, now, knows no bounds and expansion of the mental niche progresses exponentially. The sexual psyche of men is deeply flawed; shaped by primitive hunting and killing techniques over millions of years, it is not open to modification.

From Hitler in the 1930's and 40's, to today's powerfully patriarchal leaders, the male is trapped in a primitive sexual psyche. It renders him virtually powerless to self-modification, or to influences outside the zone of male dominance. Non-male attributes are seen as weak, indicators of fallibility, of failure, lack of purpose, excessive sensitivity, lack of direction.

To be influential, must women be surrogate men, adopting their rationale and personality? Are they to be led, too, into aggressive sexuality; forms of behaviour and dress that are exaggerations of femininity, directed at gaining acceptance as compliant equals?

Perhaps the days of patriarchy are coming to an end; the signs are that males themselves are losing heart. Perhaps the very systems organised to empower them have the opposite effect. Their power (as male human animals) becomes surrogate to the sophisticated tools they now employ to sustain the human patriarchy. The god of technology may soon slip out of their grasp. Then the history of previous cohorts serve as an example; diminished roles lead to loss of identity, loneliness,

241

depression, substance abuse and dependency, early death by illness or suicide.

No matter how complex and powerful are the machines onto which they project their faltering personality, unless the male sexual psyche can mature, the fate awaiting them (even, and especially, the most powerful) is of the fateful nature that signified the end of the last world war at Hiroshima. It's likely to be the end of more than just the war itself.

A time will come when patriarchy, of necessity, must cease to direct an onward march of human civilisation towards self-extinction. The future failure of the biosphere to sustain higher life forms is already apparent if there is no change.

So, if patriarchy is to die, like many of its male progeny, it may be from suicide.

CHAPTER 17

Femina Sapiens

If patriarchy has demonstrably failed, it is well past time that the other half of the human race be given the opportunity see if it can do a better job. It can't do worse.

Patriarchy does not measure up to the challenges facing modern humans. Those ideals we aspire to for our species - for humanity, justice, equality in law, safety, the right to follow a peaceful life, immunity from coercion, outlawing of rape, abuse, brutality and aggression towards others; that the list is endless and has never been achieved is the worst indictment there could be for *homo sapiens.*

Femina sapiens should, at least, be given a chance. Though 'homo' in Latin derives from the word 'earth', nevertheless the word has come to be associated with 'man'. Femina sapiens means 'wise female being', and what better ideal could there be than for this age-old creature to take her proper place at last in shaping further human evolution?

Femina sapiens will represent much more clearly the true relation between humankind, the environment and other species. She is not descended from (that is, characterised by) the hunter-killer gangs who so successfully brought us from around 5.6 million years ago (the Pliocene) to the Palaeocene, through the Palaeolithic and Neolithic, the Bronze age and Iron age. The 'modern era', arguably beginning in the 'classical era', could already be described as partly civilised, but the dominant male, trapped in the sexual psyche of earlier epochs, prompted reactions that were instinctively defensive/aggressive.

Like Hitler, the resort to force covered the weaknesses inherent in the psyche; a ruthless leader conveniently brought spoils and excess, the kind of surplus that a primitive ancestor and his tribe would greet as triumph. Pity the poor defeated; for now the conflict was habitually against his own species. Warfare raged around the world, as it had since the armies of the Middle East contested the riches of the West and vice-versa.

Yet power overshadowed the subtler interdependent relationship between species and the environment. *Homo sapiens* imagined the world created for his own use, conveniently ordained and created by a sort of super-human, much like a conquering king. The heady illusion of supremacy infects all males. Science is now their happy hunting ground, promising safe haven. Within science, linear logic guarantees the closed loop of self-consistency that constitutes proof and foregone conclusions.

The environment – the biosphere of the Earth – has been under threat most seriously since the demise of agriculture in favour of industrialisation and the 'scientific revolution'. And in this the male triumphed again; technology and invention accelerating progress to a point, today, which reveals a biosphere seriously under threat, possibly facing irremediable harm.

We humans now need to admit the fact that all species have evolved to occupy a particular niche in the biosphere. Our niche is an unnatural circumstance owing to a mutation in an early brain that led to recall, and, thence to tool-use. Despite that, because of that, we have no prior claim to exploit all the resources of the Earth; by doing so we encounter grave risks. The integrity of the biosphere is at risk.

An alternative to Patriarchy is overdue, not to be over-thrown, but ameliorated into one that takes into account the obvious risks, for there is no 'alternative' world. The love of Science for other worlds and other lands is folly; no other planet can support more than a few score of humans at unbelievable cost. Any environment cosmically close would be so dire that few would suffer it for long.

The alternative reservoir of human intelligence is right here on Earth. The ones with the insight and creativity for understanding all the fundamental issues is in our community of females. Men have suppressed female energies and talents over millennia, in favour of solutions in linear logic, to achieve direct, rapid and ruthless means to control and exploit scarce resources. The notion that raw material can be 'imported' from sources on other planets is pitiful and ludicrous. These ideas demonstrate the crude nature of the male way of thinking.

The female sexual psyche imbues 50% of the human population with a view that opposes 'head on' the continuing debasement of the Earth's natural environment. To them, such waste and profligacy is counterintuitive; the female view has a pedigree probably stretching back far beyond that of the male. Sensory intuition suggests immediately, in all circumstances, an obvious truth; the environment is small, fragile and limited. It can no longer withstand male ambitions for its exploitation.

Women are (though men resist these ideas) resourceful, intelligent, intuitive, creative and original thinkers. They are also sympathetic and empathetic to a degree beyond that of the male. There are myriad resources available to us through replenishment, a way of using resources without exhausting them. These 'soft resources' are invisible to male linear logic, which requires an identifiable final solution to signify completion, conquest, death; 'job done!'

Feminine intuition emphasizes sustainability as a means of benefiting from a resource, without its exhaustion, but rather its continuance and not 'its destruction' as 'a satisfactory and satisfying out-come'. The diversity of the biosphere is worth garnering in a way that encourages re-growth, the ability to be reborn.

Translating this into social organisation by consent, redefining objectives and method is no threat, merely a loss of those male appetites that – though not at fault of their own being the outcome of early evolutionary success - are nevertheless ultimately destructive. This is the loss of only a part of male influence, compensated for by greater opportunity for mutual co-operation. It is a positive step to relieve the male of some of the stress, under which his psyche suffers.

This is what shapes marriage; a successful strategy that benefits a species with a complex social structure. By diluting or diminishing its influence, we diminish female roles and status, casting women merely as accessories to continuing patriarchy.

Though females carry the same mutated memory and recall process, unlike the male, its influence on behaviour is probably not only modified hormonally but also by a more mature sexual psyche; that of mother, her garnering nature, as protector and nurturer. Her qualifications to share life and help shape human progress are impeccable.

Woman's 'rights' are not, or should never be, those 'rights' sanctioned by men according to the time-honoured rules of patriarchal society. Women must construct for themselves a parallel matriarchy, with its rules and customs, rights and privileges, hierarchy and governance, born of female culture and morality.

Women's degrees of cognition, precognition, recognition, the high values ascribed to their episodic memory, are as high in women as in men. There is a suspicion that, if women were able to design their own experiments, scores would be higher in every kind of anthropological dynamic. Maybe, as women co-operate better and thereby create more effective outcomes for society, they will eventually, as in a marriage, bring their skills to bear in forging a balanced matriarchal and patriarchal society.

True, there will be domestics 'spats', but Homo sapiens has a long way to go.

Evolution is evolving, still. But nature has trumped evolution with a species that has deviated from the long-accepted Darwinian model; patriarchal and primitive. Emma Darwin witnessed first-hand the pain and suffering of that old world, ordered to confine strictly both male and female. She would have recognised and welcomed the coming change. The human female is intelligent, resourceful, long-suffering, loving and intuitive.

Human-kind has need of *Femina sapiens*.

APPENDICES

" it is solely the comparison, with past ideas, which makes consciousness — & which tells one of reality"

Charles Darwin. Notebook 'M'., entry 103. Cambridge University Darwin Archive.

The notebooks in the Cambridge Archives demonstrate that Darwin was an inveterate recorder of his own thoughts, scribbling them down in 'stream of consciousness' fashion. Regrettably, he neglected to elaborate on this particular insight. It contains the kernel of a new approach to Evolutionary thinking. If he had considered the structure of memory and the   notions associated with it by later investigators, notably Tulving, a more coherent picture might have emerged, resolving the difficulties apparent in attempts to equate animals and humans and obviating much of the distress his theory caused.

A Modern View of 'Darwinism'

The Neoteric Theory of Evolution

Modern genetics shows that humans are more closely related to Chimpanzees and other apes than even Charles Darwin could have suspected. Yet in behaviour and social complexity, modern man is, by several orders of magnitude, more 'advanced'. This difference has arisen, in evolutionary terms, in an eye blink of time. Does the ability to recall events at will separate humans from other species?

Contrary to widely understood interpretations of human evolution, does this place him outside normal evolutionary pathways? Here, in showing that conscious or voluntary recall is essential for the development of tools, a distinction is apparent between *extemporized* tools (part of some animals' behavioural repertoire) and what the author names *exterpolated* tools - the result of a mental process in hominids involving recall:

"Homo sapiens, taking full advantage of its awareness of its continued existence in time, has transformed the natural world into one of culture and civilization that our distant ancestors, let alone members of other

species, possibly could not imagine". (Tulving. Annu. Rev. Psychol. 2002.53:1-25)

True tool-making (the production of *manifest* tools) is not possible without recall. Furthermore, concepts fundamental to human thought, such as 'sequence' and 'order', also depend upon recall and are absent in animals. Therefore number, language, ordered thought, all characteristics uniquely human, derive from the ability to recall events.

Mutation probably gave rise to this unique differentiator. Manifest tools in the form of modified flints, bones and antlers enabled hominids to compete with much larger predators 'catapulting' them into a role far more advanced than allowed for by the time-scales and mechanisms of natural selection as these are conventionally understood. This leads the author to conclude that humans can be classified as a 'pseudo species', that is; uniquely, they fall outside the accepted Darwinian definition of evolution.

Given these conclusions, the author proposes that the evolutionary mechanism, with the first appearance of humans, has itself changed; that evolution in the brain architecture of a hominid accelerated the adaptations necessary to produce modern humans. Because these adaptations are not provided for in classical Darwinian concepts of evolution by natural selection, but instead

251

allow humans to occupy a 'mental niche' that expands exponentially, the author tentatively names this 'new wave' of evolution 'Neoteric evolution' to emphasize the difference between humans and all other species yet classified.

Shards from flint workings have been found in the rift valley region of Africa (and cut marks on bone made by flint tools dated to 3.5 million years ago) This evidence of tool use makes it necessary to re-classify *Homo-sapiens* as arriving at a much earlier date than previously assumed. The author suggests that classification should apply to earlier hominine types demonstrating tool-making abilities and this cultural marker is more apposite in determining inclusion or exclusion of other hominids in the ancestry of *Homo sapiens.* Further, that tools of wood and other degradable material predate tools of stone and flint. Abductive reasoning suggests that, though no evidence survives for these earlier manifest tools, tools of this type, of wood leather and bone, still in use by contemporary cultures, justify this conclusion.

'Neoteric' Evolution thus extends the premises of Classical Evolutionary Theory to include recall as a precursor of human identity, distinguishing the human as a unique species that departed from the classical patterns of evolution as described by Darwin. That the niche the human occupies is now largely based upon a continuing expansion of a 'mental niche', humans may be described as a pseudo species (adopting the definition of Lalland and Odling-Smee introduced in their description of beavers).

Neoteric theory also suggests that classification of types of human ancestor might need to be revisited following new discoveries such as remains of *Homo that floresiensis;* From these finds, including elegant stone tools such as blades and scrapers, the assertion that brain size has a direct relation to intellectual ability appears misleading. *H.floresiensis* adult individuals had the stature of children, their brain cavities tiny compared to *H.habilis,* for example. Yet they certainly excelled at the production of finely crafted implements in stone, flint and other materials

Thus, the linear classification of those belonging to the genus *Homo* should realistically depend on the evidence of tool making and use, not size of skull, jaw development or lack, volume of brain cavity, etc. If adopted, the dating of 'human' activity must be pushed back to at least 3.8 million years to take account of tools produced by human ancestors at Olduvai.

Abductive evidence, supported by etymological data, suggests other tools not of stone, such as those made of woven fibre, animal skin, bark, wood, etc., though they do not survive, push back the date for signs of human cognitive emergence to perhaps another four millions years, or more.

The Electronic Niche.

The modern era is characterised by the dominance of electronic messaging - the internet. With the invention of the use of electronic pulses via wireless or charged cables, man - or rather men - revolutionised the mental niche, accelerated its expansion a hundred- or a thousand-fold. Growing by the minute (or the second) this recent rapid expansion of the human mental function (transposing thought, word and deed into binary data) signifies another step- change in evolution.

All life is affected. The expansion of the mental niche feeds back into every aspect of life within the limited confines of the Earth's biosphere. Animal, plant marine life and micro-organisms are all affected.

Men may have employed sophisticated linear logic to create, diversify and expand electronics into every known (and some unknown) forms of media and communication, but it is women who respond most vigorously to its influence and function; immediately, tirelessly and with a kind of commitment lacking in the (male) originators.

This is how the Pew Research Centre puts it: women, they suggest, are "biologically wired" for social networking. Hard evidence for sensory intuition?

The nature of sensory intuition has not been defined. It may never be. This is as hard a task of attempting to isolated in the human brain the site of recall. Evidently, it is not possible to 'pin down' - as the linear-logic mind would wish - the site of recall or episodic memory; it seems the brain 'works', not like a 'mechanism'; the 'brain wave' is not simply a metaphor. Constructs really do appear to be more like a wave at sea than the tick of a clock or the whirring of wound-up springs. The 'wave' passes though the water, air or space leaving it untouched, unaltered. The shimmering, wavy illumination we detect via the electrodes used in brain research hints at the significance of the phenomenon of cognition. The matter itself does not. The cells involved have to be 'firing' in order to function as 'ghosts in the machine'.

We don't know why or how the brain of Kim Peek- a good portion of is volume of the brain matter entirely absent - is able to memorise entire telephone directories and compose symphonies. Mr Peek may have problems tying his shoelaces, but so what? Slip-ons are just as good.

We know that the men who created the web hoped it would speed communication between distant brains. According to Brandwatch, men are more likely to cite ' business reasons' for using the web, or to " build influence, perform research and gather relevant contacts".

But women, especially with social media, are the prime movers, overall 76% are women 74% are men. For the 'biologically wired', "networking" has expanded from the doorstep to an electronically ordered society (Pew).

How likely, one survey asked, is the respondent likely to 'stay in touch'?

69% of women, 54% of men.

Broken down into sectors or devices:

Mobile: 69% of women   39% of men

Facebook :76% of women   66% of men

Average 'posts' " 394 by women   254 by men

Message time (mins) 10   women   > 7   men

Among internet users, a greater percentage are women; 76%: (Twitter, Facebook,Tumblr, Pinterest, Snapchat, Instagram).Women represent the bulk of the business which largely male inventors and male-orientated merchandisers and investors now exploit.

This pattern of social and business interaction may present challenges to women. A widening divide (both social and economic) evolves between smaller number of male entrepreneurs who manipulate an increasing number of women. Patriarchy is now enshrining itself in modern society through electronic innovation.

Power is accruing again - like the priests of ancient religions - to a small caste of males. Future power lies within the expanding niche of the electronic network - 'the net'.

Sadly, males will miss out more than females. The niche is almost second nature to females in the way it works; massaging sensory intuition, confirming its vital role in the female psyche. Meanwhile, the male need for a physical role within which to 'enshrine' his individual identity is diminishing daily. Male suicide figures were on the rise (slightly) until recently, though these have always been higher significantly than for females.

Currently male suicide figures are down from 16.8 to 16.6 deaths per 100,000 in 2016, females up from 5.2 to 5.4 (Office for National Statistics). A steady decline for both sexes has occurred since 1988 (21.2 deaths, per 100,000 then for males and 8.3 for females).

The electronic age is growing at an exponential rate; more data feeds into the mental niche and, in turn, influences us as we resort more and more to non-physical 'in-put'. What outcomes may occur is hard to judge, but a flattening curve on the charts for suicide leads to speculation that physical disassociation from 'real life' might presage return to earlier decades.

Alternatively, growing awareness among greater numbers of people may 'give us pause'. Dawning awareness of the fragile environment we live in may lead to a growing confidence in dealing with 'those ills we have'.

NOTES: